Marine Capture Fisheries

Marine Capture Fisheries

Dr. Sunita Rai

RANDOM PUBLICATIONS
NEW DELHI (INDIA)

Marine Capture Fisheries

ISBN 978-93-5111-516-8

Published in 2015 in India by

RANDOM PUBLICATIONS

Reprint, 2017

4376-A/4B, Gali Murari Lal, Ansari Road
New Delhi-110 002
Phone : +9111-43580356, 011-23289044, 011-43142548
e-mail: sales@randompublications.com,
info@randompublications.com, randomexports@gmail.com

Type Setting by : Friends Media, Delhi-110089
Digitally Printed at : Replika Press Pvt. Ltd.

Preface

Fisheries and aquaculture make crucial contributions to the world's well-being and prosperity. In the last five decades, world fish food supply has outpaced global population growth, and today fish constitutes an important source of nutritious food and animal protein for much of the world's population. In addition, the sector provides livelihoods and income, both directly and indirectly, for a significant share of the world's population.

Close to 90 per cent of the world's fishery catches come from oceans and seas, as opposed to inland waters. These marine catches have remained relatively stable since the mid-nineties. Most marine fisheries are based near the coast. This is not only because harvesting from relatively shallow waters is easier than in the open ocean, but also because fish are much more abundant near the coastal shelf, due to the abundance of nutrients available there from coastal upwelling and land runoff.

Marine capture fisheries resources are usually considered close to full exploitation worldwide with about half of them fully exploited, one quarter over exploited, depleted or recovering from depletion and one quarter only with some capacity to produce more than they presently do. Capture fisheries resources are usually exploited and managed on a stock-by-stock basis. Stocks present a wide range of characteristics that affect the fisheries exploiting them: their mono- or multi-species composition, size, value and distribution.

This book provides a comprehensive overview of the current trends and status of the marine capture fisheries sector. It discusses the issues of marine fisheries governance and the trends towards ecosystem-based management. The book will be of interest to students, fisheries scientists, fishermen and policy makers involved with fisheries management.

Editor

Contents

	Preface	**v**
1.	**Status of World Fisheries**	**1**
	Fishery Industry	2
	Historical Development of Fisheries	4
	Status and Trends	9
	Capture Fisheries Production	30
	Global Aquaculture Production	35
	Food Fish Production	37
	Status of Fishery Resources	48
2.	**Marine Fishery Resources**	**58**
	Large Marine Ecosystems	60
	Production and Trends in Marine Fisheries	70
	Marine Catch Composition	71
	Catch Fluctuations	72
	Exploitation of Fishery Resources	74
3.	**Methods of Fish Capture**	**77**
	Types of Capture Fisheries	77
	Fish Capture Technology	83
	Fishing Gears and Methods	84
	Exclusive Economic Zones (EEZ)	104
	Destructive Fishing Practices	106
4.	**Administration of Capture Fisheries**	**111**

	Fisheries Monitoring, Control and Surveillance (MCS)	111
	Components of MCS	117
	Design Considerations	121
	Infrastructure Requirements	130
	Impact of New Technology	138
	Fisheries Legislation	141
	Management Measures	144
	Fisheries Management Plans	155
	Towards Successful MCS Operations	161
	Implementing MCS Plans	166
	Operational Procedures	167
5.	**Impact of Climate Change on Capture Fisheries**	**172**
	Fisheries' Contribution to Economic Development	172
	Exposure and Sensitivity of Fisheries to Climate Change	173
	Fisheries Categories	175
	Vulnerability and Resilience	176
	Socio-economic Context of Fisheries	177
	Climate Change and Climate Variability	178
	Units and Scales of Analysis	178
	Fisheries and Climate Change Mitigation	179
	Impacts by Sector	183
	Market and Trade Impacts	187
	Potential Positive Impacts	187
	Observed and Future Impacts	188
	Vulnerability of Regions, Groups and Hot Spots	190
	Adaptation of Fisheries to Climate Change	195
6.	**Ecosystem Approach to Capture Fisheries**	**203**
	Impacts from Fisheries on the Environment	204
	What is an Ecosystem Approach to Fisheries?	212
	Principles and Concepts of EAF	215
	Making EAF operational	216

Moving towards EAF management 217
Data and Information Requirements 224
Management Measures and Approaches 228
Threats to Implementing EAF 241
7. Code of Conduct for Responsible Fisheries 245
Article 1. Nature and Scope of the Code 247
Article 2. Objectives of the Code 248
Article 3. Relationship with Other International Instruments 249
Article 4. Implementation Monitoring and Updating 249
Article 5. Special Requirements of Developing Countries 250
Article 6. General Principles 250
Article 7. Fisheries Management 254
Article 8. Fishing Operations 262
Article 9. Aquaculture Development 269
Article 10. Integration of Fisheries into Coastal Area Management 272
Article 11. Post-harvest Practices and Trade 273
Article 12. Fisheries Research 277
Annex 1. Background to the Origin and Elaboration of the Code 280
Annex 2 . Resolution 285
Bibliography 288
Index 290

1

Status of World Fisheries

Generally, a fishery is an entity engaged in raising or harvesting fish which is determined by some authority to be a fishery. According to the FAO, a fishery is typically defined in terms of the "people involved, species or type of fish, area of water or seabed, method of fishing, class of boats, purpose of the activities or a combination of the foregoing features". The definition often includes a combination of fish and fishers in a region, the latter fishing for similar species with similar gear types.

A fishery may involve the capture of wild fish or raising fish through fish farming or aquaculture. Directly or indirectly, the livelihood of over 500 million people in developing countries depends on fisheries and aquaculture. Overfishing, including the taking of fish beyond sustainable levels, is reducing fish stocks and employment in many world regions.

Fisheries are harvested for their value. They can be saltwater or freshwater, wild or farmed. Examples are the salmon fishery of Alaska, the cod fishery off the Lofoten islands, the tuna fishery of the Eastern Pacific, or the shrimp farm fisheries in China. Capture fisheries can be broadly classified as industrial scale, small-scale or artisanal, and recreational.

Close to 90% of the world's fishery catches come from oceans and seas, as opposed to inland waters. These marine catches have remained relatively stable since the mid-nineties. Most marine fisheries are based near the coast. This is not only because harvesting from relatively shallow waters is easier than in the open ocean, but also because fish are much more abundant near the coastal shelf, due to the abundance of nutrients available there from coastal upwelling and land runoff. However, productive wild fisheries also exist in open oceans, particularly by seamounts, and inland in lakes and rivers.

Most fisheries are wild fisheries, but farmed fisheries are increasing. Farming can occur in coastal areas, such as with oyster farms, but more typically occur inland, in lakes, ponds, tanks and other enclosures.

There are species fisheries worldwide for finfish, mollusks, crustaceans and echinoderms, and by extension, aquatic plants such as kelp. However, a very small number of species support the majority of the world's fisheries. Some of these species are herring, cod, anchovy, tuna, flounder, mullet, squid, shrimp, salmon, crab, lobster, oyster and scallops. All except these last four provided a worldwide catch of well over a million tonnes in 1999, with herring and sardines together providing a harvest of over 22 million metric tons in 1999. Many other species are harvested in smaller numbers.

According to the Food and Agriculture Organization (FAO), the world harvest in 2005 consisted of 93.3 million tonnes captured by commercial fishing in wild fisheries, plus 48.1 million tonnes produced by fish farms. In addition, 1.3 million tons of aquatic plants (seaweed etc.) were captured in wild fisheries and 14.8 million tons were produced by aquaculture. The number of individual fish caught in the wild has been estimated at 0.97-2.7 trillion per year (not counting fish farms or marine invertebrates).

Fishery Industry

The fishery industry is involved in many activities covering all aspects of the business spectrum: buying inputs, applying for loans, paying for labour, planning for the future and making profits. These exist throughout a fisheries system: in capture, processing and marketing. Also, as in any industry, artisanal or industrial fisheries have characteristics which must be identified in order to better understand the fishery industry overall. The world's fisheries have a common characteristic; that is, the existence of industrial fisheries alongside artisanal fisheries. Developing countries have both types of fisheries, with a mix and a relative importance, that should be studied case-by-case, and in some countries even region-by-region.

Fisheries development means an evolution from the subsistence to artisanal stage, then to the small-scale and finally to the semi-industrial and industrial stages. This development usually implies external investment and/

or reinvestment and a redistribution of human resources. Development also means avoiding obsolescence, and keeping abreast of new technologies, procedures, market requirements, etc. A characteristic of the dynamics of sustainable development is the need for an endless chain of techno-economic decisions. In both developed and developing countries static fisheries systems are found, unchanged and stable for centuries. These issues are not discussed here since many other factors may contribute to them.

Whereas industrial fisheries could be identified according to the technologies utilised and investment, there is no universally-accepted definition of small-scale fisheries. Existing fisheries are classified into several groups: artisanal and industrial or commercial; small and large scale; operating range of vessels; or types of nets used (Thailand), size of vessels (Indonesia, Philippines) or distance from shore or a combination of the three (Malaysia). What one country considers large-scale, is often regarded as small-scale by another. It is generally accepted though, that large-scale fisheries operate in deep waters, whereas small-scale fisheries have limited catch capacity, operate in an adverse socio-economic environment, are confined to a narrow strip of sea and land, and are utilised by a community which has limited options and is dependent on local resources.

For those outside the fishery industry, and sometimes even for those conversant with it, the endeavour of fish production, utilisation and marketing often appears rather strange, difficult to understand and generally obscure. This is partly due to the intrinsic complexity of fisheries that embraces the latest technology to obtain food from the wild, to the use of satellites, electronic equipment, robots and bioengineering. It is also partly due to the large number of fish species marketed, which makes fish a very different protein source from, for instance, meat and chicken.

Part of the complexity stems from the large number of products available, processes and their variations, and catching methods ranging from beach nets and canoes to high-tech factory trawlers. The complexity at local, regional and international levels is also related to deep historical and cultural reasons, without which consumption habits, ad hoc laws and regulations, fishermen's behaviour, political moves and the commoner's position regarding fish and fisheries are very difficult to understand. Whereas in some industries history and culture are practically irrelevant to their economics, in fisheries they usually play an important part although very often overlooked or taken for granted.

Historical Development of Fisheries

Since prehistory, fish has been caught and eaten first by hominids (Australopithicus and Homo erectus) and then by men (Homo sapiens). There is enough archaeological evidence that men were already catching fish in the Lower Palaeolithic Age, more than 100 000 years ago, and the earliest record of fish as food for Homo sapiens is 380 000 years old. In more recent prehistoric times, there is plenty of evidence that European populations habitually utilised fish as food, salmon being one of the most widely consumed fish, and some Amerindian and African populations were known shellfish gatherers.

Fish was appreciated by ancient Egyptian and Chinese civilisations. The first recorded recipe (a fish salad based on marinated and spiced carp) is from ancient China dated 1300 B.C. The ancient Romans' passion for fish is very well known; the highest recorded price for an auctioned fish (two live red mullets) ever paid has not yet been surpassed. Fresh fish has always been preferred. The Chinese have been trading live fish for over 3,000 years. In the Roman Empire, the best quality fish was also kept, transported and sold live. There is evidence that the Chinese utilised natural ice to keep fresh fish about 1000 B.C.. Ancient Romans also used ice mixed with seaweed to keep fresh fish.

Fish drying, smoking and salting were used to cure fish from very ancient times in different cultures. Fish salting and fish fermentation were already a flourishing integrated industry almost in contemporary terms (capture, farming, processing, packaging, transport and distribution) in the Roman Empire about 100 B.C. Curing techniques have been revised and refined several times during the history of mankind and are still widely used. It is reported that salting herring onboard was introduced by the Dutch in the fourteenth century. This allowed longer fishing trips and reduced post-harvest losses, improving the production and economics of salted herring. Likewise, in the twentieth century, freezing trawlers and factory vessels were introduced to freeze and process fish onboard.

Fish farming became an affirmed technology in China between 2000 and 1500 B.C. and has never ceased to be a source of food. The first treaty on carp culture was written around 475 B.C. by a Chinese named Fan Li. This contains useful advice on design, construction, harvesting and economic management of fish ponds.

Aquaculture in Italy can be traced back to the fifth-sixth century B.C., when the Etruscans developed it in coastal lagoons of the Tyrrhenian Sea (Mediterranean). Near the Port of Cosa, they performed impressive engineering works, cutting a channel in the rock (tagliata di Ansedonia) for the hydraulic management of a lagoon of between 500 and 1000 ha.

Later, the ancient Romans, as demonstrated by archaeological studies developed in the same place, about 100 B.C.1 a sophisticated brackishwater aquaculture complex, integrating it with fish processing, packaging, and shipment of final products to many places in the Mediterranean basin. This complex also included seasonal coastal fishing of species like mackerel and tuna. The remainder of the Etruscan coastal lagoon of Cosa, the Burano lagoon, is still exploited.

The inhabitants of Cosa incorporated at a later stage the exploitation of the Orbetello lagoon (2,700 ha), one of the early examples of efficient fisheries management of coastal lagoons. The aquaculture activities in this area have been and still are integrated with marine fisheries. The case of Cosa, as many other examples that can be found, particularly in the Mediterranean, China and Japan, testimony the tight relationship between self-sustained fishery activities, social and cultural aspects and environment.

There is also enough historical evidence that aquaculture in Europe continued to be an important source of protein for almost a thousand years after the collapse of the Roman Empire. Fish production in Europe was always a mixture of cultured and wild fish from marine and inland water sources. A traveller at the beginning of the eighteenth century observed that fish consumption in Poland consisted of salted herring, salted-dried cod, marine fish from the Baltic Sea and farmed fish in the hinterland. Fish consumption has been affected positively and negatively by religions, taboos, political decisions and beliefs throughout history, which in turn have affected fish production and marketing.

Fish consumption during the Middle Ages in Europe was promoted by the Catholic Church which ordered 166 days of fasting a year during which fish could be eaten. This situation was usually reinforced by rulers; for instance, Charlemagne ordered that all his farms have fish ponds. Alternatively, the Reformation in England reduced the number of fishing vessels, severely affected freshwater fisheries, and nearly abolished aquaculture. This situation was to change again in the seventeenth century with the development of the herring fishery.

The importance of aquaculture in Europe declined for many reasons which varied from country to country; aquaculture collapsed in Germany during the Thirty Years' War and did not recover until the end of the nineteenth century. The development of marine fisheries during the nineteenth and twentieth centuries reduced even further the relative contribution of aquaculture; a change mainly due to the large yield of marine captures in respect of investment, and the worldwide development of markets for certain types of salted fish.

The importance of marine fishery captures, particularly cod and herring, increased in Europe from the thirteenth and fourteenth centuries, particularly in the northern countries. The fourteenth century was a period of famine in Europe and probably obliged coastal populations to increase the pressure on marine fishery resources. The plague that devastated Europe between 1347 and 1351 affected the coastal populations that consumed fish as a source of protein and lipids far less than the inland populations which were stressed and prone to illness due to starvation.

During the thirteenth century, herring shifted its spawning migrations from the North Sea to the Baltic Sea. It gave the Hanseatic League the possibility of developing the largest fishery industry of the time, replacing Denmark which had previously exploited this resource. This situation lasted until the sixteenth century when herring returned to spawn again in the North Sea and this time the Netherlands grasped the opportunity to develop as a major fishing country.

This return of the herring to the North Sea also triggered off the development of the Scottish herring fishery during the seventeenth century. By the second half of the seventeenth century salted fish was one of the main British exports to Europe. In the period January March 1665 at the port of Leghorn 9,020 barrels of salted herring were unloaded, 345 barrels of salted salmon and 500 small barrels of salted pressed sardine transported from Britain, and comprising almost the total cargo of eleven vessels. It could be presumed that the access to markets for salted herring was a main cause for friction between Britain and the Netherlands at that time.

Fishing rights were very often part of peace treaties between European countries. The Treaty of Utrech, although recognising, in principle, to Spanish fishermen, the right to fish for cod and whales off Newfoundland, in practice it deprived them such a right. This forced the Spaniards to look for alternatives, for instance, to increase fishing and salting sardine off

Galicia and to increase fishery imports. As one of the measures to tackle the lack of fish the Spanish King Charles IV founded, in 1789, a fishing company based in Puerto Deseado with the purpose of fishing and salting. The Spaniards returned to fish off Newfoundland only in 1921. The development of this conflict is still making news.

By the early nineteenth century, new methods were needed to extend the shelf-life of fish and fishery products. Canning of fish and meat were invented by the Frenchman, Appert as a means of supplying food to Napoleon's army. The effect of low temperature on the keeping time and quality of fresh fish was known throughout the ages. "Weather freezing" is a method applied since time immemorial by Eskimos; it consists in leaving fish outside in windy conditions of subfreezing temperatures and was probably the start of the modern fish-freezing industry. There are records that indicate that "weather freezing" was carried out in the region of the Great Lakes early in the nineteenth century and the method was still applied there in the 1960s. It was not until the development of mechanical refrigeration in the nineteenth century that ice and cold facilities became readily available. The French engineer, F. Carré, constructed the first ice-block machine, presented at the Great Exhibition of London in 1859.

In 1877 the first cargo of mutton, using mechanical refrigeration, was transported from France to Buenos Aires (Argentina) onboard the vessel "Le Frigorifique", mainly to demonstrate that intercontinental transport of frozen foods was feasible. The Argentineans, with the assistance of Carré, fitted out a second vessel "Paraguay" and started to transport frozen meat to Europe. The Americans were the first to realise the potential market possibilities of frozen fish, and by 1865 they started to freeze fish by putting it in pans surrounded by ice and salt. Around 1880 ammonia refrigeration machines had begun to be utilised and by the end of the nineteenth century fish freezing was an important industry in the USA which had started to export frozen salmon to Europe. At that time, fish was being frozen in Europe, but in smaller quantities than in the USA.

The initial quality of frozen fishery products was very poor and the process not well understood. In 1929, an American, Clarence Birdseye, decided to return to the source and find out why Eskimo frozen fish was of much better quality than mechanical frozen fish. After spending time with the Eskimos of Labrador, he found that the secret was in the freezing speed; in the USA he developed the first plate and double-belt freezers and initiated the era of "quick freezing".

After the development of the herring and cod fisheries, most of the marine fish consumed in Europe came from sailing boats, like the trawling smacks that drifted downwind with beam trawls. Steam propulsion was introduced in European fishing boats towards the end of the nineteenth century and these mechanised boats replaced sailing fishing boats.

Internal combustion engines were introduced in European fisheries at the turn of the nineteenth and twentieth centuries, and completely replaced steam propulsion by the 1960s. After the second world war, fishing boats in Europe started to introduce echo-sounders and echo-ranging to detect fish, and these methods were taken up by industrial fishing fleets all over the world. The first full-scale factory stern trawler, "Fairtry", was constructed in 1953 in Aberdeen, Scotland. It incorporated plate freezers and a combination of plate and blast freezers. Following years of research and development at Torry Research Station, Aberdeen, Scotland, in 1961 commercial vertical plate freezers were installed for the first time onboard the stern trawler "Lord Nelson". Since then there has been a rapid development of fish processing and freezing at sea.

Two important developments during the 1950s and 1960s that enhanced the advantage of mechanisation of fishing vessels were the introduction of synthetic fibre in net construction, and the rational design of fishing nets. These developments revamped old fishing methods like trawling, gillnetting and longlining and dramatically increased the catches. However, this development was not conducted pari passu with proper management of existing resources and led to significant overexploitation and economic losses being incurred by the world fisheries. These losses have resulted from fleet overcapacity and overinvestment; it is estimated that 46% of the landed value of total world catches were required as return on capital invested in the fleet and this was disproportionately high.

The problems of quality and adulteration of food became apparent during the late part nineteenth century, both in Europe and the USA, and the lack of scientific and technical information regarding food, in particular fish, became clear. Problems of food wastage observed during the first world war, and probably the growing problems associated with the food provision to larger urban populations, convinced Governments to create ad hoc research and development institutions. The first research institute dealing with fish technology was set up in Norway in 1892.

Freezing and the possibility of developing reliable cold chains lead to the introduction of a large number of products in fish markets in developed countries, particularly after the second world war. Among these were products which made the specific fish species less important and introduced the concept of prepared and semi-prepared products such as fishfingers, fish-sticks, fish-burgers and fish-pastes.

There are many type of fish-pastes. However, the type that has taken the lead is that originally known as "kamaboko" in Japan. "Kamaboko" and derived products were a traditional fishery product mainly consumed in Japan but also known in China and other Asian countries. It was originally produced artisanally from a few fresh fish; in 1960 the Japanese started to produce kamaboko from frozen kneaded fish flesh from Alaska pollack and mechanised the whole production process.

Production of kamaboko and kamaboko-based products escalated in Japan during the 1970s, the technology began to spread and was utilised in the rest of the world. In or around 1975 in Japan, and later in other developed countries, kamaboko began to be utilised for the production of analogs, and opened up the possibility of utilisation of the less exploited species and potentially the reduction of the post-harvest losses.

It is clear that there has been, since the second half of the nineteenth century, a constant reduction in the successive "short-term" horizons of the fishery industry, as a result of the introduction of technical and scientific innovations, and market changes. It is extremely important for developing countries to understand this reduction of the "short-term" fishery industry structure in order to achieve self-sustainability and not waste investment. The reduction in the "short-term" horizons makes it necessary to consider the medium and long-term horizons for proper self-sustained development.

Status and Trends

Capture fisheries and aquaculture supplied the world with about 148 million tonnes of fish in 2010 (with a total value of US$217.5 billion), of which about 128 million tonnes was utilized as food for people, and preliminary data for 2011 indicate increased production of 154 million tonnes, of which 131 million tonnes was destined as food. With sustained growth in fish production and improved distribution channels, world fish food supply has grown dramatically in the last five decades, with an average growth rate of 3.2 percent per year in the period 1961–2009, outpacing the increase of 1.7

percent per year in the world's population. World per capita food fish supply increased from an average of 9.9 kg (live weight equivalent) in the 1960s to 18.4 kg in 2009, and preliminary estimates for 2010 point to a further increase in fish consumption to 18.6 kg. Of the 126 million tonnes available for human consumption in 2009, fish consumption was lowest in Africa (9.1 million tonnes, with 9.1 kg per capita), while Asia accounted for two-thirds of total consumption, with 85.4 million tonnes (20.7 kg per capita), of which 42.8 million tonnes was consumed outside China (15.4 kg per capita). The corresponding per capita fish consumption figures for Oceania, North America, Europe, and Latin America and the Caribbean were 24.6 kg, 24.1 kg, 22.0 kg and 9.9 kg, respectively. Although annual per capita consumption of fishery products has grown steadily in developing regions (from 5.2 kg in 1961 to 17.0 kg in 2009) and in low-income food-deficit countries (LIFDCs, from 4.9 kg in 1961 to 10.1 kg in 2009), it is still considerably lower than in more developed regions, although the gap is narrowing. A sizeable share of fish consumed in developed countries consists of imports, and, owing to steady demand and declining domestic fishery production (down 10 percent in the period 2000–2010), their dependence on imports, in particular from developing countries, is projected to grow in coming years.

China has been responsible for most of the increase in world per capita fish consumption, owing to the substantial increase in its fish production, particularly from aquaculture, despite a downward revision of China's production statistics for recent years. China's share in world fish production grew from 7 percent in 1961 to 35 percent in 2010. Driven by growing domestic income and an increase in the diversity of fish available, per capita fish consumption in China has also increased dramatically, 31.9 kg in 2009, with an average annual rate of 6.0 percent in the period 1990–2009. If China is excluded, annual fish supply to the rest of the world in 2009 was about 15.4 kg per person, higher than the average values of the 1960s (11.5 kg), 1970s (13.5 kg), 1980s (14.1 kg) and 1990s (13.5 kg).

Fish and fishery products represent a very valuable source of protein and essential micronutrients for balanced nutrition and good health. In 2009, fish accounted for 16.6 percent of the world population's intake of animal protein and 6.5 percent of all protein consumed. Globally, fish provides about 3.0 billion people with almost 20 percent of their intake of animal protein, and 4.3 billion people with about 15 percent of such protein. Differences among developed and developing countries are apparent in the contribution of fish to animal protein intake. Despite the relatively lower levels of fish consumption in developing countries, the share contributed by fish was significant at about 19.2 percent, and for LIFDCs it was 24.0 percent. However, in both developing and developed countries, this share has declined slightly in recent years as consumption of other animal proteins has grown more rapidly.

Overall global capture fisheries production continues to remain stable at about 90 million tonnes although there have been some marked changes in catch trends by country, fishing area and species. In the last seven years (2004–2010), landings of all marine species except anchoveta only ranged between 72.1 million and 73.3 million tonnes. In contrast, the most dramatic changes, as usual, have been for anchoveta catches in the Southeast Pacific, which decreased from 10.7 million tonnes in 2004 to 4.2 million tonnes in 2010. A marked decrease in anchoveta catches by Peru in 2010 was largely a result of management measures (e.g. fishing closures) applied to protect the high number of juveniles present as a consequence of the La Niña event (cold water). This action paid dividends in 2011 when anchoveta catches exceeded their 2009 level. Inland water capture production continued to grow continuously, with an overall increase of 2.6 million tonnes in the period 2004–2010.

The Northwest Pacific is still by far the most productive fishing area. Catch peaks in the Northwest Atlantic, Northeast Atlantic and Northeast Pacific temperate fishing areas were reached many years ago, and total production had declined continuously from the early and mid-2000s, but in 2010 this trend was reversed in all three areas. As for mainly tropical areas, total catches grew in the Western and Eastern Indian Ocean and in the Western Central Pacific. In contrast, the 2010 production in the Western Central Atlantic decreased, with a reduction in United States catches by about 100 000 tonnes, probably mostly attributable to the oil spill in the Gulf of Mexico. Since 1978, the Eastern Central Pacific has shown a series of fluctuations in capture production with a cycle of about 5–9 years. The latest peak was in 2009, and a declining phase may have started in 2010. Both the Mediterranean–Black Sea and the Southwest Atlantic have seen declining catches, with decreases of 15 and 30 percent, respectively, since 2007. In the Southeast Pacific (excluding anchoveta) and the Southeast Atlantic, both areas where upwelling phenomena occur with strongly varied intensity each year, historical catch trends have been downward in both areas.

In the Eastern Central Atlantic, production has increased in the last three years, but there are some reporting inconsistencies for this area.

Chilean jack mackerel catches have declined for this transboundary resource with a very wide distribution in the South Pacific, ranging from the national exclusive economic zones (EEZs) to the high seas. After having peaked at about 5 million tonnes in the mid-1990s, catches were about 2 million tonnes in the mid-2000s but have since declined abruptly, and the 2010 catches were 0.7 million tonnes, the lowest level since 1976. In contrast, Atlantic cod catches have increased by almost 200 000 tonnes in the last two years. In fact, in 2010, the whole group of gadiform species (cods, hakes, haddocks, etc.) reversed the negative trend of the previous three years in which it had declined by 2 million tonnes. Preliminary data for this group also report growing catches for 2011. Capture production of other important commercial species groups such as tunas and shrimps remained stable in 2010. The highly variable catches of cephalopods resumed growth after a decrease in 2009 of about 0.8 million tonnes. In the Antarctic areas, interest in fishing for krill resumed, and a catch increase of more than 70 percent was registered in 2010.

Total global capture production in inland waters has increased dramatically since the mid-2000s with reported and estimated total production at 11.2 million tonnes in 2010, an increase of 30 percent since 2004. Despite this growth, it may be that capture production in inland waters is seriously underestimated in some regions. Nevertheless, inland waters are considered as being overfished in many parts of the world, and human pressure and changes in the environmental conditions have seriously degraded important bodies of freshwater (e.g. the Aral Sea and Lake Chad). Moreover, in several countries that are important in terms of inland waters fishing (e.g. China), a good portion of inland catches comes from waterbodies that are artificially restocked. It is not clear to what extent improvements in the statistical coverage and stock enhancement activities may be contributing to the apparent increase in inland fishery production. Growth in the global inland water catch is wholly attributable to Asian countries. With the remarkable increases reported for 2010 production by India, China and Myanmar, Asia's share is approaching 70 percent of global production. Inland water capture production in the other continents shows different trends. Uganda and the United Republic of Tanzania, fishing mostly in the African Great Lakes, and Nigeria and Egypt, with river fisheries, remain the main producers in Africa. Catches in several South and North American countries have been reported as shrinking. Increased European production between 2004 and 2010 is all attributable to a rise of almost 50 percent in catches of the Russian Federation. Inland fishery production is marginal in countries in Oceania.

In the last three decades (1980–2010), world food fish production of aquaculture has expanded by almost 12 times, at an average annual rate of 8.8 percent. Global aquaculture production has continued to grow, albeit more slowly than in the 1980s and 1990s. World aquaculture production attained another all-time high in 2010, at 60 million tonnes (excluding aquatic plants and non-food products), with an estimated total value of US$119 billion. When farmed aquatic plants and non-food products are included, world aquaculture production in 2010 was 79 million tonnes, worth US$125 billion. About 600 aquatic species are raised in captivity in about 190 countries for production in farming systems of varying input intensities and technological sophistication. These include hatcheries producing seeds for stocking to the wild, particularly in inland waters.

In 2010, global production of farmed food fish was 59.9 million tonnes, up by 7.5 percent from 55.7 million tonnes in 2009 (32.4 million tonnes in 2000). Farmed food fish include finfishes, crustaceans, molluscs, amphibians (frogs), aquatic reptiles (except crocodiles) and other aquatic animals (such as sea cucumbers, sea urchins, sea squirts and jellyfishes), which are indicated as fish throughout this document. The reported grow-out production from aquaculture is almost entirely destined for human consumption. The total farmgate value of food fish production from aquaculture is estimated at US$119.4 billion for 2010.

Aquaculture production is vulnerable to adverse impacts of disease and environmental conditions. Disease outbreaks in recent years have affected farmed Atlantic salmon in Chile, oysters in Europe, and marine shrimp farming in several countries in Asia, South America and Africa, resulting in partial or sometimes total loss of production. In 2010, aquaculture in China suffered production losses of 1.7 million tonnes caused by natural disasters, diseases and pollution. Disease outbreaks virtually wiped out marine shrimp farming production in Mozambique in 2011.

The global distribution of aquaculture production across the regions and countries of different economic development levels remains imbalanced. In 2010, the top ten producing countries accounted for 87.6 percent by quantity and 81.9 percent by value of the world's farmed food fish. Asia accounted for 89 percent of world aquaculture production by volume in 2010, and this was dominated by the contribution of China, which accounted for more than 60 percent of global aquaculture production volume in 2010. Other major producers in Asia are India, Viet Nam, Indonesia, Bangladesh, Thailand, Myanmar, the Philippines and Japan. In Asia, the share of freshwater aquaculture has been gradually increasing, up to 65.6 percent in 2010 from around 60 percent in the 1990s. In terms of volume, Asian aquaculture is dominated by finfishes (64.6 percent), followed by molluscs (24.2 percent), crustaceans (9.7 percent) and miscellaneous species (1.5 percent). The share of non-fed species farmed in Asia was 35 percent (18.6 million tonnes) in 2010 compared with 50 percent in 1980.

In North America, aquaculture has ceased expanding in recent years, but in South America it has shown strong and continuous growth, particularly in Brazil and Peru. In terms of volume, aquaculture in North and South America is dominated by finfishes (57.9 percent), crustaceans (21.7 percent) and molluscs (20.4 percent). In Europe, the share of production from brackish and marine waters increased from 55.6 percent in 1990 to 81.5 percent in 2010, driven by marine cage culture of Atlantic salmon and other species. Several important producers in Europe have recently ceased expanding or have even contracted, particularly in the marine bivalve sector. In 2010, finfishes accounted for three-quarters of all European aquaculture production, and molluscs one-quarter. Africa has increased its contribution to global production from 1.2 percent to 2.2 percent in the past ten years, mainly as a result of rapid development in freshwater fish farming in sub-Saharan Africa. African aquaculture production is overwhelmingly dominated by finfishes, with only a small fraction from marine shrimps and marine molluscs. Oceania accounts for a minor share of global aquaculture production and this consists mainly of marine molluscs and finfishes, with the latter increasing owing mainly to the development of farming of Atlantic salmon in Australia and chinook salmon in New Zealand.

The least-developed countries (LDCs), mostly in sub-Saharan Africa and in Asia, remain minor in terms of their share of world aquaculture production (4.1 percent by quantity and 3.6 percent by value) with the main producers including Bangladesh, Myanmar, Uganda, the Lao People's Democratic Republic and Cambodia. However, some developing countries in Asia and the Pacific (Myanmar and Papua New Guinea), sub-Saharan

Africa (Nigeria, Uganda, kenya, Zambia and Ghana) and South America (Ecuador, Peru and Brazil) have made rapid progress to become significant or major aquaculture producers in their regions. In contrast, in 2010, developed industrialized countries produced collectively 6.9 percent (4.1 million tonnes) by quantity and 14 percent (US$16.6 billion) by value of the world's farmed food fish production, compared with 21.9 percent and 32.4 percent, respectively, in 1990. Aquaculture production has contracted or stagnated in Japan, the United States of America and several European countries. An exception is Norway, where, thanks to the farming of Atlantic salmon in marine cages, aquaculture production grew from 151 000 tonnes in 1990 to more than one million tonnes in 2010.

Freshwater fishes dominate global aquaculture production (56.4 percent, 33.7 million tonnes), followed by molluscs (23.6 percent, 14.2 million tonnes), crustaceans (9.6 percent, 5.7 million tonnes), diadromous fishes (6.0 percent, 3.6 million tonnes), marine fishes (3.1 percent, 1.8 million tonnes) and other aquatic animals (1.4 percent, 814 300 tonnes). While feed is generally perceived to be a major constraint to aquaculture development, one-third of all farmed food fish production (20 million tonnes) is currently achieved without artificial feeding, as is the case for bivalves and filter-feeding carps. However, the percentage of non-fed species in world production has declined gradually from more than 50 percent in 1980 to the present level of 33.3 percent, reflecting the relatively faster body-growth rates achieved in the culture of fed species and increasing consumer demand for higher trophic-level species of fishes and crustaceans.

Fisheries and aquaculture provided livelihoods and income for an estimated 54.8 million people engaged in the primary sector of fish production in 2010, of whom an estimated 7 million were occasional fishers and fish farmers. Asia accounts for more than 87 percent of the world total with China alone having almost 14 million people (26 percent of the world total) engaged as fishers and fish farmers. Asia is followed by Africa (more than 7 percent), and Latin America and the Caribbean (3.6 percent). About 16.6 million people (about 30 percent of the world total) were engaged in fish farming, and they were even more concentrated in Asia (97 percent), followed by Latin America and the Caribbean (1.5 percent), and Africa (about 1 percent). Employment in the fisheries and aquaculture primary sector has continued to grow faster than employment in agriculture, so that by 2010 it represented 4.2 percent of the 1.3 billion people economically active in the broad agriculture sector worldwide, compared with 2.7 percent in 1990. In the last five years, the number of people engaged in fish farming has increased by 5.5 percent per year compared with only 0.8 percent per year for those in capture fisheries, although capture fisheries still accounted for 70 percent of the combined total in 2010. It is apparent that, in the most important fishing nations, the share of employment in capture fisheries is stagnating or decreasing while aquaculture is providing increased opportunities. Europe experienced the largest decrease in the number of people engaged in capture fishing, with a 2 percent average annual decline between 2000 and 2010, and almost no increase in people employed in fish farming. In contrast, Africa showed the highest annual increase (5.9 percent) in the number of people engaged in fish farming in the same period, followed by Asia (4.8 percent), and Latin America and the Caribbean (2.6 percent). Overall, production per person is lower in capture fisheries than in aquaculture, with global outputs of 2.3 and 3.6 tonnes per person per year respectively, reflecting the huge numbers of fishers engaged in small-scale fisheries.

Apart from the primary production sector, fisheries and aquaculture provide numerous jobs in ancillary activities such as processing, packaging, marketing and distribution, manufacturing of fish-processing equipment, net and gear making, ice production and supply, boat construction and maintenance, research and administration. All of this employment, together with dependants, is estimated to support the livelihoods of 660–820 million people, or about 10–12 percent of the world's population.

The total number of fishing vessels in the world in 2010 is estimated at about 4.36 million, which is similar to previous estimates. Of these, 3.23 million vessels (74 percent) are considered to operate in marine waters, with the remaining 1.13 million vessels operating in inland waters. Overall, Asia has the largest fleet, comprising 3.18 million vessels and accounting for 73 percent of the world total, followed by Africa (11 percent), Latin America and the Caribbean (8 percent), North America (3 percent) and Europe (3 percent). Globally, 60 percent of fishing vessels were engine-powered in 2010, but although 69 percent of vessels operating in marine waters were motorized, the figure was only 36 percent for inland waters. For the fleet operating in marine waters, there were also large variations among regions, with non-motorized vessels accounting for less than 7 percent of the total in Europe and the Near East, but up to 61 percent in Africa.

Over 85 percent of the motorized fishing vessels in the world are less than 12 m in length overall (LOA). Such vessels dominate in all regions, but markedly so in the Near East, and Latin America and the Caribbean. About 2 percent of all motorized fishing vessels corresponded to industrialized fishing vessels of 24 m and larger (with a gross tonnage [GT] of roughly more than 100 GT) and that fraction was larger in the Pacific and Oceania region, Europe, and North America.

Data from some countries indicate a recent expansion in their fleets. For example, the motorized fishing fleets in Malaysia, Cambodia and Indonesia increased by 26, 19 and 11 percent, respectively, between 2007 and 2009, and Viet Nam reported a 10 percent increase in offshore fishing vessels (those with engines of more than 90 hp) between 2008 and 2010. The case of Sri Lanka illustrates potential overshoot in efforts to re-establish a fishing fleet, of which 44 percent of the motorized vessels were destroyed by the tsunami that swept the region at the end of 2004, with the result that by 2010 there were 11 percent more motorized vessels than before the tsunami.

Many countries have policies to reduce overcapacity in their fishing fleets. China's marine fishing vessel reduction plan for 2003–2010 did achieve a reduction by 2008 close to the target, but since then both the number of vessels and total combined power have started to increase again. Japan implemented various schemes that resulted in a net reduction of 9 percent in the number of vessels, but a net increase of 5 percent in combined power between 2005 and 2009. The evolution in the combined number, tonnage, and power of European Union fishing vessels indicates a downward tendency in the last decade and the combined EU-15 motorized fishing fleet achieved a net reduction of 8 percent in the number of vessels and of 11 percent in power between 2005 and 2010. Other important fishing nations that achieved a net reduction in fleet size in the period 2005–2010 include Iceland, Norway and the Republic of korea.

The world's marine fisheries increased markedly from 16.8 million tonnes in 1950 to a peak of 86.4 million tonnes in 1996, and then declined before stabilizing at about 80 million tonnes. Global recorded production was 77.4 million tonnes in 2010. The Northwest Pacific had the highest production with 20.9 million tonnes (27 percent of the global marine catch) in 2010, followed by the Western Central Pacific with 11.7 million tonnes (15 percent), the Northeast Atlantic with 8.7 million tonnes (11 percent), and the Southeast Pacific, with a total catch of 7.8 million tonnes (10 percent). The proportion of non-fully exploited stocks has decreased gradually since 1974 when the first FAO assessment was completed. In contrast, the percentage of overexploited stocks has increased, especially in the late 1970s and 1980s, from 10 percent in 1974 to 26 percent in 1989. After 1990, the number of overexploited stocks continued to increase, albeit at a slower rate. Increases in production from these overexploited stocks may be possible if effective rebuilding plans are put in place. The fraction of fully exploited stocks, which produce catches that are very close to their maximum sustainable production and have no room for further expansion and require effective management to avoid decline, has shown the smallest change over time, with its percentage stable at about 50 percent from 1974 to 1985, then falling to 43 percent in 1989 before gradually increasing to 57 percent in 2009. About 29.9 percent of stocks are overexploited, producing lower yields than their biological and ecological potential and in need of strict management plans to restore their full and sustainable productivity in accordance with the Johannesburg Plan of Implementation that resulted from the World Summit on Sustainable Development, which demands that all overexploited stocks be restored to the level that can produce maximum sustainable yield by 2015, a target that seems unlikely to be met. The remaining 12.7 percent of stocks were non-fully exploited in 2009, and these are under relatively low fishing pressure and have some potential to increase their production although they often do not have a high production potential and require proper management plans to ensure that any increase in the exploitation rate does not result in further overfishing.

Most of the stocks of the top ten species, which account in total for about 30 percent of world marine capture fisheries production, are fully exploited and, therefore, have no potential for increases in production, while some stocks are overexploited and increases in their production may be possible if effective rebuilding plans are put in place. The two main stocks of anchoveta in the Southeast Pacific, Alaska pollock in the North Pacific and blue whiting in the Atlantic are fully exploited. Atlantic herring stocks are fully exploited in both the Northeast and Northwest Atlantic. Japanese anchovy in the Northwest Pacific and Chilean jack mackerel in the Southeast Pacific are considered to be overexploited. Chub mackerel stocks are fully exploited in the Eastern Pacific and the Northwest Pacific. The largehead hairtail was estimated in 2009 to be overexploited in the main fishing area in the Northwest Pacific.

Among the seven principal tuna species, one-third were estimated to be overexploited, 37.5 percent were fully exploited, and 29 percent non-fully exploited in 2009. Although skipjack tuna continued its increasing trend up to 2009, further expansion should be closely monitored, as it may negatively affect bigeye and yellowfin tunas (multispecies fisheries). In the long term, the status of tuna stocks (and consequently catches) may further deteriorate unless there are significant improvements in their management. This is because of the substantial demand for tuna and the significant overcapacity of tuna fishing fleets. Concern about the poor status of some bluefin stocks and the inability of some tuna management organizations to manage these stocks effectively led to a proposal in 2010 to ban the international trade in Atlantic bluefin tuna under the Convention on International Trade in Endangered Species of Wild Fauna and Flora (CITES) and, although the proposal was ultimately rejected, the concern remains.

The declining global marine catch over the last few years together with the increased percentage of overexploited fish stocks and the decreased proportion of non-fully exploited species around the world convey the strong message that the state of world marine fisheries is worsening and has had a negative impact on fishery production. Overexploitation not only causes negative ecological consequences, but it also reduces fish production, which further leads to negative social and economic consequences. To increase the contribution of marine fisheries to the food security, economies and well-being of the coastal communities, effective management plans must be put in place to rebuild overexploited stocks. The situation seems more critical for some highly migratory, straddling and other fishery resources that are exploited solely or partially in the high seas. The United Nations Fish Stocks Agreement that entered into force in 2001 should be used as a legal basis for management measures of the high seas fisheries.

In spite of the worrisome global situation of marine capture fisheries, good progress is being made in reducing exploitation rates and restoring overexploited fish stocks and marine ecosystems through effective management actions in some areas. In the United States of America, 67 percent of all stocks are now being sustainably harvested, while only 17 percent are still overexploited. In New Zealand, 69 percent of stocks are above management targets, reflecting mandatory rebuilding plans for all fisheries that are still below target thresholds. Similarly, Australia reports overfishing for only 12 percent of stocks in 2009. Since the 1990s, the

Newfoundland–Labrador Shelf, the Northeast United States Shelf, the Southern Australian Shelf, and California Current ecosystems have shown substantial declines in fishing pressure such that they are now at or below the modelled exploitation rate that gives the multispecies maximum sustainable yield of the ecosystem. These and other successes can serve as examples to assist in more effective management of other fisheries.

The information summarizing the state of the major marine fish stocks is impossible to duplicate for the state of most of the world's inland fisheries, for which the exploitation rate is often not the main driver affecting the state of the stocks. Other drivers such as habitat quantity and quality, aquaculture in the form of stocking and competition for freshwater, influence the state of the majority of inland fishery resources much more than exploitation rates do. Water abstraction and diversion, hydroelectric development, draining wetlands, and siltation and erosion from land-use patterns can negatively affect inland fishery resources regardless of the rate of exploitation. Conversely, stock enhancement from aquaculture facilities, which is widely practised in inland waters, can keep catch rates high in the face of increased fishing and in spite of an ecosystem that is not capable of producing that level of catch through natural processes. Overexploitation also affects inland fishery resources, but the result is generally a change in species composition and not necessarily a reduced overall catch. Catches are often higher where smaller and shorter-lived species become the main component of the catch; however, the smaller fish may be much less valuable. Another issue complicating the assessment of inland fishery resources is the definition of a "stock". Very few inland fisheries have stocks that are defined precisely or are defined at the level of species. There are notable exceptions such as the Lake Victoria Nile perch and Tonle Sap dai fisheries, but many inland fishery resources are defined by watershed or river and comprise numerous species. Taking all of these considerations into account, FAO is leading efforts to improve data collection and develop new assessment methodologies for inland fishery resources that are so important but often underestimated in terms of their economic, social and nutritional benefits and contribution to livelihoods and food security. The intention is to utilize the new methodology to provide a more robust and informative summary of the state of the world's inland capture fishery resources in the future.

Concerning utilization of the world's fish production, 40.5 percent (60.2 million tonnes) was marketed in live, fresh or chilled forms, 45.9 percent

(68.1 million tonnes) was processed in frozen, cured or otherwise prepared forms for direct human consumption, and 13.6 percent destined for non-food uses in 2010. Since the early 1990s, there has been an increasing trend in the proportion of fisheries production used for direct human consumption rather than for other purposes. Whereas in the 1980s about 68 percent of the fish produced was destined for human consumption, this share increased to more than 86 percent in 2010, equalling 128.3 million tonnes. In 2010, 20.2 million tonnes was destined to non-food purposes, of which 75 percent (15 million tonnes) was reduced to fishmeal and fish oil; the remaining 5.1 million tonnes was largely utilized as fish for ornamental purposes, for culture (fingerlings, fry, etc.), for bait, for pharmaceutical uses as well as for direct feeding in aquaculture, for livestock and for fur animals. Of the fish destined for direct human consumption, the most important product form was live, fresh or chilled fish, with a share of 46.9 percent in 2010, followed by frozen fish (29.3 percent), prepared or preserved fish and cured fish (9.8 percent). Freezing represents the main method of processing fish for human consumption, and it accounted for 55.2 percent of total processed fish for human consumption and 25.3 percent of total fish production in 2010.

The proportion of frozen fish grew from 33.2 percent of total production for human consumption in 1970 to reach a record high of 52.1 percent in 2010. The share of prepared and preserved forms remained rather stable during the same period, and it was 26.9 percent in 2010. Developing countries have experienced a growth in the share of frozen products (24.1 percent of the total fish for human consumption in 2010, up from 18.9 percent in 2000) and of prepared or preserved forms (11.0 percent in 2010, compared with 7.8 percent in 2000). Owing to deficiencies in infrastructure and processing facilities, together with well-established consumer habits, fish in developing countries is commercialized mainly in live or fresh form (representing 56.0 percent of fish destined for human consumption in 2010) soon after landing or harvesting. Cured forms (dried, smoked or fermented) still remain a traditional method to retail and consume fish in developing countries, although their share in total fish for human consumption is declining (10.9 percent in 2000 compared with 8.9 percent in 2010). In developed countries, the bulk of production destined for human consumption is commercialized frozen or in prepared or preserved forms.

Fishmeal is produced from whole fish or fish remains resulting from processing. Small pelagic species, in particular anchoveta, are the main contributors for reduction, and the volume of fishmeal and fish oil produced worldwide fluctuates annually according to the fluctuations in the catches of these species, which are strongly influenced by the El Niño phenomenon. Fishmeal production peaked in 1994 at 30.2 million tonnes (live weight equivalent) and has followed a fluctuating trend since then. In 2010, it dropped to 15.0 million tonnes owing to reduced catches of anchoveta, representing a decrease of 12.9 percent compared with 2009, of 18.2 percent compared with 2008, and of 42.8 percent with respect to 2000. Waste from commercial fish species used for human consumption is increasingly used in feed markets, and a growing percentage of fishmeal is being obtained from trimmings and other residues from the preparation of fish fillets. About 36 percent of world fishmeal production was obtained from offal in 2010.

Technological development in food processing and packaging is progressing rapidly. Processors of traditional products have been losing market share as a result of long-term shifts in consumer preferences as well as in processing and in the general fisheries industry. Processing is becoming more intensive, geographically concentrated, vertically integrated and linked with global supply chains. These changes reflect the increasing globalization of the fisheries value chain, with large retailers controlling the growth of international distribution channels. The increasing practice of outsourcing processing at the regional and world levels is very significant, but further outsourcing of production to developing countries might be restricted by sanitary and hygiene requirements that are difficult to meet as well as by growing labour costs. At the same time, processors are frequently becoming more integrated with producers, especially for groundfish, where large processors in Asia, in part, rely on their own fleet of fishing vessels. In aquaculture, large producers of farmed salmon, catfish and shrimp have established advanced centralized processing plants. Processors that operate without the purchasing or sourcing power of strong brands are also experiencing increasing problems linked to the scarcity of domestic raw material, and they are being forced to import fish for their business.

Fish and fishery products continue to be among the most traded food commodities worldwide, accounting for about 10 percent of total agricultural exports and 1 percent of world merchandise trade in value terms. The share of total fishery production exported in the form of various food and feed items increased from 25 percent in 1976 to about 38 percent (57 million tonnes) in 2010. In the same period, world trade in fish and fishery products grew significantly also in value terms, rising from US$8 billion to US$102 billion. Sustained demand, trade liberalization policies, globalization of food systems and technological innovations have furthered the overall increase in international fish trade. In 2009, reflecting the general economic contraction affecting consumer confidence in major markets, trade dropped by 6 percent compared with 2008 in value terms as a consequence of falling prices and margins, whereas traded volumes, expressed in live weight equivalent, increased by 1 percent to 55.7 million tonnes. In 2010, trade rebounded strongly, reaching about US$109 billion, with an increase of 13 percent in value terms and 2 percent in volume compared with 2009. The difference between the growth in value and volume reflects the higher fish prices experienced in 2010 as well as a decrease in the production of and trade in fishmeal. In 2011, despite the economic instability experienced in many of the world's leading economies, increasing prices and strong demand in developing countries pushed trade volumes and values to the highest level ever reported and, despite some softening in the second half of the year, preliminary estimates indicate that exports exceeded US$125 billion.

Since late 2011 and early 2012, the world economy has entered a difficult phase characterized by significant downside risks and fragility, and key markets for fisheries trade have slowed sharply. Among the factors that might influence the sustainability and growth of fishery trade are the evolution of production and transportation costs and the prices of fishery products and alternative commodities, including meat and feeds. In the last few decades, the growth in aquaculture production has contributed significantly to increased consumption and commercialization of species that were once primarily wild-caught, with a consequent price decrease, particularly in the 1990s and early 2000, with average unit values of aquaculture production and trade declining in real terms. Subsequently, owing to increased costs and continuous high demand, prices have started to rise again. In the next decade, with aquaculture accounting for a much larger share of total fish supply, the price swings of aquaculture products could have a significant impact on price formation in the sector overall, possibly leading to more volatility.

As for trade, fish prices also contracted in 2009 but have since rebounded. The FAO Fish Price Index indicates that average prices in 2009 declined by 7 percent compared with 2008, then increased by 9 percent in 2010 and by more than 12 percent in 2011. Prices for species from capture fisheries increased by more than those for farmed species because of the larger impact from higher energy prices on fishing vessel operations than on farmed species. Since 2002, China has been by far the leading fish exporter, contributing almost 12 percent of 2010 world exports of fish and fishery products, or about US$13.3 billion, and increasing further to US$17.1 billion in 2011. A growing share of fishery exports consists of reprocessed imported raw material. Thailand has established itself as a processing centre of excellence largely dependent on imported raw material, while Viet Nam has a growing domestic resource base and imports only limited, albeit growing, volumes of raw material. Viet Nam has experienced significant growth in its exports of fish and fish products, up from US$1.5 billion in 2000 to US$5.1 billion in 2010, when it became the fourth-largest exporter in the world. In 2011, its exports rose further to US$6.2 billion, linked mainly to its flourishing aquaculture industry. In 2010, developing countries confirmed their fundamental importance as suppliers to world markets with more than 50 percent of all fishery exports in value terms and more than 60 percent in quantity (live weight). For many developing nations, fish trade represents a significant source of foreign currency earnings in addition to the sector's important role as a generator of income, source of employment, and provider of food security and nutrition. The fishery industries of developing countries rely heavily on developed countries, not only as outlets for their exports, but also as suppliers of their imports for local consumption or for their processing industries. In 2010, in value terms, 67 percent of the fishery exports of developing countries were directed to developed countries. A growing share of these exports consisted of processed fishery products prepared from imports of raw fish to be used for further processing and re-export. In 2010, in value terms, 39 percent of the imports of fish and fishery products by developing countries originated from developed countries.

World imports of fish and fish products set a new record at US$111.8 billion in 2010, up 12 percent on the previous year and up 86 percent with respect to 2000. Preliminary data for 2011 point to further growth, with a 15 percent increase. The United States of America and Japan are the major importers of fish and fishery products and are highly dependent on imports for about 60 percent and 54 percent, respectively, of their fishery consumption. China, the world's largest fish producer and exporter, has significantly increased its fishery imports, partly a result of outsourcing, as

Chinese processors import raw material from all major regions, including South and North America and Europe, for re-processing and export. Imports are also being fuelled by robust domestic demand for species not available from local sources, and, in 2011, China became the third-largest importer in the world. The European Union is by far the largest single market for imported fish and fishery products owing to its growing domestic consumption. However, it is extremely heterogeneous, with markedly different conditions from country to country. European Union fishery imports reached US$44.6 billion in 2010, up 10 percent from 2009, and representing 40 percent of total world imports. However, if intraregional trade is excluded, the European Union imported fish and fishery products worth US$23.7 billion from suppliers outside the European Union, an increase of 11 percent from 2009. In addition to the major importing countries, a number of emerging markets have become of growing importance to the world's exporters. Prominent among these there are Brazil, Mexico, the Russian Federation, Egypt, Asia and the Near East in general. In 2010, developed countries were responsible for 76 percent of the total import value of fish and fishery products, a decline compared with the 86 percent of 1990 and 83 percent of 2000. In terms of volume (live weight equivalent), the share of developed countries is significantly less, 58 percent, reflecting the higher unit value of products imported by developed countries.

Owing to the high perishability of fish and fishery products, 90 percent of trade in fish and fishery products in quantity terms (live weight equivalent) consists of processed products. Fish are increasingly traded as frozen food (39 percent of the total quantity in 2010, compared with 25 percent in 1980). In the last four decades, prepared and preserved fish have nearly doubled their share in total quantity, going from 9 percent in 1980 to 16 percent in 2010. However, trade in live, fresh and chilled fish represented 10 percent of world fish trade in 2010, up from 7 percent in 1980, reflecting improved logistics and increased demand for unprocessed fish. Trade in live fish also includes ornamental fish, which is high in value terms but almost negligible in terms of quantity traded. In 2010, 71 percent of the quantity of fish and fishery products exported consisted of products destined for human consumption. The US$109 billion exports of fish and fishery products in 2010 do not include an additional US$1.3 billion for aquatic plants (62 percent), inedible fish waste (31 percent) and sponges and corals (7 percent). In the last two decades, trade in aquatic plants has increased significantly, rising from US$0.2 billion in 1990 to US$0.5 billion in 2000 and to US$0.8 billion in 2010, with China as the major exporter and Japan as the leading importer.

A recent major event related to governance of fisheries and aquaculture has been the UN Conference on Sustainable Development, known as Rio+20, to renew political commitment for sustainable development, assess progress and gaps in the implementation of existing commitments, and address new challenges. The two themes of the conference were the institutional framework for sustainable development and the support of a green economy. As a concept, the green economy aims to ensure that resource exploitation contributes to sustainability, inclusive social development and economic growth, while seeking to counter the notion that sustainability and growth are mutually exclusive.

At Rio+20, FAO promoted the message that there will be no green economy without sustainable growth in agriculture (including fisheries) and that improved management and efficiencies throughout the food value chain can increase food security while using fewer natural resources. The message calls for policies that create incentives to adopt sustainable practices and behaviour and promotes the wide application of ecosystem approaches. FAO also contributed to interagency submissions to Rio+20 concerning the sustainable management of the world's oceans with a focus on the green economy as it relates to marine and coastal resources, sustainable use and poverty eradication, small-scale fisheries and aquaculture operations, and the potential contribution of small island developing States.

The dependence of the fisheries and aquaculture sectors on ecosystem services means that supporting sustainable fishing and fish farming can provide incentives for wider ecosystem stewardship. The greening of fisheries and aquaculture requires recognition of their wider societal roles within a comprehensive governance framework. There are several mechanisms to facilitate this transition, including adopting an ecosystem approach to fisheries and aquaculture with fair and responsible tenure systems to turn resource users into resource stewards.

Small-scale fisheries employ more than 90 percent of the world's capture fishers, and their importance to food security, poverty alleviation and poverty prevention is becoming increasingly appreciated. However, the lack of institutional capacity and the failure to include the sector in national and regional development policies hamper their potential contribution. Since

2003, the FAO Committee on Fisheries (COFI) has promoted efforts to improve the profile of, and understand the challenges and opportunities facing, small-scale fishing communities in inland and marine waters. It has also recommended the development of international voluntary guidelines to complement the Code of Conduct for Responsible Fisheries (the Code) as well as other international instruments with similar purposes. The preparation of the guidelines is expected to contribute to policy development and have considerable impact on securing small-scale fisheries and creating benefits, especially in terms of food security and poverty reduction. The guidelines promote good governance, including transparency and accountability, participation and inclusiveness, social responsibility and solidarity, a human rights approach to development, gender equality, and respect and involvement of all stakeholders.

Regional fishery bodies (RFBs) are the primary organizational mechanism through which States work together to ensure the long-term sustainability of shared fishery resources. The term RFB also embraces regional fisheries management organizations (RFMOs), which have the competence to establish binding conservation and management measures. As intergovernmental organizations, RFBs depend on the political will of their member Governments to implement agreed measures and undertake reform. Most RFBs are experiencing difficulties in fulfilling their mandates (many of which are outdated). However, important progress in extending the global coverage of RFBs is being made through new, strengthened and emerging bodies. In addition, numerous RFBs have been undergoing independent reviews of their performance. The 2010 United Nations Review Conference described the modernizing of RFMOs as a priority and noted that progress had been made in developing best practices for RFMOs and in reviewing their performance against emerging standards. Ten RFBs have so far undergone performance reviews. The Review Conference observed that performance reviews were generally recognized as being useful, particularly when they led to the adoption of new management measures.

Illegal, unreported and unregulated (IUU) fishing and related activities (often encouraged by corrupt practices) threaten efforts to secure long-term sustainable fisheries and promote healthier and more robust ecosystems. The international community continues to express its grave concern at the extent and effects of IUU fishing. Developing countries, often with limited technical capacity, bear the brunt of this IUU fishing, which undermines their limited efforts to manage fisheries, denies them revenue and adversely affects their attempts to promote food security, eradicate poverty and achieve sustainable livelihoods. However, there are indications that IUU fishing is moderating in some areas (e.g. the Northeast Atlantic Ocean) as policies and measures take effect.

Nonetheless, the international community is deeply frustrated by the failure of many flag States to meet their primary responsibilities under international law, which are to exercise effective control over their fishing vessels and ensure compliance with conservation and management measures. Of particular concern are those vessels flying flags of "non-compliance", which are flags belonging to States that are either unable or unwilling to exercise effective control over their vessels. As a result, the burden of controlling these rogue vessels is gradually falling on coastal States, port States, RFBs and others. This has led FAO Members to request that a Technical Consultation on Flag State Performance be convened. It is anticipated that the outcome will be a set of voluntary criteria for assessing the performance of flag States together with a list of possible actions to be taken against vessels flying the flags of States not meeting such criteria and possibly an agreed procedure for assessing compliance.

Although their achievements in terms of limiting IUU fishing vary widely, most RFBs promote and implement measures to combat IUU fishing. The measures range from more passive activities such as awareness building and dissemination of information (mainly RFBs without fisheries management functions) to aggressive port, air and surface surveillance programmes (RFMOs).

Beyond national boundaries, there is increasing need for international cooperation to improve global fisheries management of shared marine resources and to preserve the associated employment and other economic benefits of sustainable fisheries. Recognizing this, the European Union and the United States of America, as leaders in the global fish trade, undertook (in 2011) to cooperate bilaterally to combat IUU fishing by keeping illegally caught fish out of the world market. Strengthening fisheries management capacity is fundamental in developing countries in order to facilitate sustainable fisheries and to reduce the impacts of IUU fishing. Capacity development is especially important to support the full and effective implementation of existing and new global instruments such as the 2009 Port State Measures Agreement to combat IUU fishing.

Governance of aquaculture has become increasingly important and has made remarkable progress. To improve planning and policy development in aquaculture, many Governments utilize the Code as well as FAO guidelines and manuals on farming techniques promoted by industry organizations and development agencies. Several countries have adequate national aquaculture development policies, strategies, plans and laws, and use "best management practices". The insert: 2011 FAO Technical Guidelines on Aquaculture Certification constitute an additional important tool for good governance of the sector. By setting minimum substantive criteria for developing aquaculture certification standards, these guidelines provide direction for the development, organization and implementation of credible aquaculture certification schemes towards orderly and sustainable development of the sector. Long-term prosperity requires technological soundness, economic viability, environmental integrity and social licence, which, in combination, also ensure that ecological well-being is compatible with human well-being.

An important component of human well-being is employment, which in aquaculture has grown rapidly in the last three decades. More than 100 million people now depend on the sector for a living, either as employees in the producing and support sectors or as their dependants. In many places, these employment opportunities have enabled young people to stay in their communities and have strengthened the economic viability of isolated areas, often enhancing the status of women in developing countries, where more than 80 percent of aquaculture output occurs. Aquaculture has been heavily promoted in several countries with fiscal and monetary incentives and this has improved accessibility to food for many households and increased aquaculture's contribution towards the Millennium Development Goals (MDGs). However, the sector has developed at a time of growing scrutiny fromthe public, improved communications and vociferous opposition groups. Although opposition groups can act as environmental and social watchdogs, putting pressure on businesses to increase transparency and improve working conditions, it is also important to consider the benefits accruing from the sector, including those related to employment.

Unfair employment practices in aquaculture, including exploitation of local labour, gender discrimination and child employment, can undermine trust in the sector, threaten the credibility of policy-makers and jeopardize markets for farmed seafood. Most countries have legislation to protect workers but compliance therewith can deter enterprises, with some opting to operate in countries with lower labour and social standards where they can gain a competitive advantage. A possible result is that Governments will be under pressure from companies to reduce labour and social standards.

Employment in aquaculture must be equitable and non-exploitative, with principled values guiding activities to induce beyond-compliance behaviour. With an ethos of corporate social responsibility, aquaculture companies would assist local communities, employ fair labour practices and demonstrate transparency. Increasingly, with rising consumer awareness, it makes good business sense for aquaculture enterprises to demonstrate that they meet the best standards. Legislation should protect labour and reflect concepts of social justice and human rights, but it needs to strike a balance as overly cumbersome regulations can make an otherwise viable business unprofitable.

Capture Fisheries Production

Overall global capture fisheries production, as derived from the FAO capture database, continues to remain stable. This does not mean that there are no changes in catch trends by country, fishing area or species, which indeed do vary significantly throughout the years, but rather that the summation of all the annual fluctuations has been close to zero in recent years.

To analyse trends, global production can be separated into three major components: marine catches excluding anchoveta (*Engraulis ringens*); anchoveta catches; and inland water catches. In the last seven years (2004–2010) forwhich detailed catch statistics are available, absolute variations in comparison with the previous year of total marine catches excluding anchoveta never exceeded 1.2 percent, ranging between 72.1 and 73.3 million tonnes. However, anchoveta catches decreased from 10.7 million tonnes in 2004 to 4.2 million tonnes in 2010, and the variation on the previous year exceeded 30 percent in two cases. In the same period, inland water capture production grew continuously, with an overall increase of 2.6 million tonnes.

A marked decrease in anchoveta catches by Peru in 2010 was mostly due to management measures (e.g. fishing closures) that were applied in the final quarter to protect the high number of juveniles present in the anchoveta stock as a consequence of the La Niña event (cold water), which had favoured spawning and generated a good recruitment. Thanks to this precautionary management decision, the 2011 anchoveta catches exceeded their 2009 level. Other preliminary reports from important fishing countries (e.g. the Russian Federation) show that 2011 should have been a year of increased catches. However, Japanese fishery production will probably have dropped significantly as the five prefectures hit by the earthquake and tsunami of 11 March 2011 accounted for about 21 percent of Japan's total marine fisheries and aquaculture production. Overall, preliminary information suggests that the total 2011 global catch should exceed 90 million tonnes, marking a return to 2006–07 levels.

Notwithstanding the protracted global economic downturn, which has reduced the funds available to national administrations, the submission rates of 2009 and 2010 catch data to FAO have remained reasonably stable. However, it is well known that the quality of fishery data is very uneven among countries. An evaluation of data quality in capture statistics submitted to FAO found that more than half of the countries reported inadequately. This percentage was greater for developing countries, but also about one-fourth of reports by developed countries were not satisfactory. Countries that should improve their data collection and reporting systems are mainly found in Africa, Asia and among the island States in Oceania and the Caribbean.

World Marine Capture Fisheries Production

With the great decrease in anchoveta catches, Peru is no longer second after China in the ranking of the major marine producer countries in terms of quantity as it has been surpassed by Indonesia and the United States of America. Some major Asian fishing countries (i.e. China, India, Indonesia, Myanmar and Viet Nam) reported significant increases in 2010, but also other countries (i.e. Norway, the Russian Federation and Spain) fishing in other areas and with more robust data collection systems showed growing catches after some years of sluggish production.

In particular, catches reported by the Russian Federation have grown by more than one million tonnes since the low point of 2004. According to the authorities of the Russian Federation, the recent increase is also a consequence of the management decision to remove excessive formalities on documentation of landing operations, as up until early 2010 landings by vessels of the Russian Federation in national ports were treated as imports. Moreover, an official forecast of the Russian Federation indicates further catch increases to a level of 6 million tonnes in 2020, representing an increase of more than 40 percent above present levels.

Besides decreased production by Peru and Chile as a consequence of the drop in anchoveta catches, other major fishing countries with downward trends in total marine catches in 2009 and 2010 were: Japan, the Republic of korea, and Thailand in Asia; Argentina, Canada and Mexico in the Americas; Iceland in Europe; and to a lesser extent New Zealand. Despite variable trends, Morocco, South Africa and Senegal maintained their positions as the three major marine producers in Africa.

The Northwest Pacific is still by far the most productive fishing area. Catch peaks in the Northwest Atlantic, Northeast Atlantic and Northeast Pacific temperate fishing areas were reached many years ago (in 1968, 1976 and 1987, respectively) and total production had declined continuously from the early and mid-2000s, but in 2010 this trend was reversed in all three areas.

As for mainly tropical areas, total catches grew in the Western and Eastern Indian Ocean and in the Western Central Pacific, and, in the last two, 2010 marked a new maximum. In contrast, the 2010 production in the Western Central Atlantic decreased, driven by the reduction in United States catches by about 100 000 tonnes, probably mostly attributable to the oil spill in the Gulf of Mexico. Since 1978, the Eastern Central Pacific has shown a series of fluctuations in capture production with a cycle of about 5–9 years. The latest peak was in 2009, and a declining phase may have started in 2010.

Both the Mediterranean–Black Sea and the Southwest Atlantic seem to be areas where fisheries are in trouble as, since 2007, total catches have decreased by 15 and 30 percent, respectively. In the two areas along the southwest sides of America and Africa, upwelling phenomena occur, although their intensity varies strongly each year. In 2010, catches in the Southeast Pacific (excluding anchoveta) decreased whereas in the Southeast Atlantic they grew, but examination of historical trends from an earlier period reveals clear downward trajectories in both areas.

Finally, in the Eastern Central Atlantic, production has increased in the last three years. However, in this area, total capture production is significantly influenced by the activities of distant-water fleets and whether their catches are reported only by the flag States or also complemented with information by some costal countries that register foreign fleet catches in their EEZ but only make these data available to FAO intermittently.

As noted above, annual catches by fishing area, country and in particular by species very often fluctuate considerably, but all these variations combined seem to have a counterbalancing effect on the global total. A demonstration of this is that catches of more than 60 percent of the species varied by more than 10 percent in comparison with 2009 but the global total (excluding anchoveta) changed by only 1.2 percent.

It is well documented that fish populations show large fluctuations in abundance, also in the absence of fishing. Although the causes are well known for some species (e.g. anchoveta – driven by changing environmental regimes), they remain unknown for many others. Besides fishes, such variations also occur in other commercial groups of species. For example, Argentina started industrial-level exploitation of *Pleoticus muelleri*, a high-value shrimp, in the 1980s. However, this species showed a major drop in 2005. Facing much reduced catches, the national authorities implemented management plans to help the species to recover. After six years, catches had rebounded tenfold reaching a new maximum recorded level in 2011.

Despite the decreased 2010 catches, anchoveta is again the most-caught species. However, also in the presence of future favourable environmental regimes, yearly catches of this species should not attain the past peaks as the Government of Peru has introduced an annual quota for the whole country, subdivided by vessel, with the purpose of stabilizing the capacity of both the fleet and processing plants.

In the list of top ten species, the most evident change is the disappearance from the list of the Chilean jack mackerel (*Trachurus murphyi*), which had been sixth in 2008. This species is a transboundary resource with a very wide distribution in the South Pacific, ranging from the national EEZs to the high seas. After having peaked at about 5 million tonnes in the mid-1990s, catches were about 2 million tonnes in the mid-2000s but have since declined abruptly, and the 2010 catches were 0.7 million tonnes, the lowest level since 1976. Atlantic cod (*Gadus morhua*) has returned to the list, with a total increase of almost 200 000 tonnes in the last two years to rank tenth in 2010, a position not reached since 1998. In fact, in 2010, the whole group of gadiform species (cods, hakes, haddocks, etc.) reversed the negative trend of the previous three years in which it had declined by 2 million tonnes. Preliminary data for this group also report growing catches for 2011.

Capture production of other important commercial species groups such as tunas and shrimps remained stable in 2010. The highly variable catches of cephalopods resumed growth after a decrease in 2009 of about 0.8 million tonnes. In the Antarctic areas, interest in fishing for krill resumed and a catch increase of more than 70 percent was registered in 2010.

Of the four marine bivalve groups, clams and cockles, which in the early 1990s contributed more than half of the overall bivalve catches, have recently accelerated their rate of decline. In 2009–2010, they were largely surpassed by scallops, which in contrast have shown a rising trend since the late 1990s. Capture production of mussels and oysters, for which reporting countries often have difficulty in separating harvest of natural populations from aquaculture production, has not varied much over the years, but an overall downward trend can be noted.

World Inland Capture Fisheries Production

Total global capture production in inland waters has increased dramatically since the mid-2000s. Total production, as submitted by countries and as estimated by FAO in cases of non-reporting, amounted to 11.2 million tonnes in 2010, an increase of 30 percent since 2004. Despite this growth, there are still claims that global production is much greater as some studies have pointed out that capture production in inland waters is seriously underestimated in some regions. However, the little well-documented evidence available concerns a limited number of countries. On the other hand, inland waters are considered as being overfished in many parts of the world, and human pressure and changes in the environmental conditions have seriously degraded important bodies of freshwater (e.g. the Aral Sea, and Lake Chad). Moreover, in several countries that are important in terms of inland waters fishing (e.g. China), a good portion of inland catches comes from waterbodies that are artificially restocked and closely monitored and, hence, it is probable that production is recorded quite carefully. Therefore, both improvements in the statistical coverage and stock enhancement activities may be contributing to the apparent increase in inland fishery production.

A closer look at the statistics shows that the growth in the global inland water catch is wholly attributable to Asian countries. With the remarkable increases reported for 2010 production by India (up 0.54 million tonnes on 2009) and by China and Myanmar (up 0.1 million tonnes each), Asia's share is approaching 70 percent of global production. Considerable increases by some major Asian countries have seriously influenced the global total in recent years but, in some cases, they seem to beconsequences of a tendency to report continuously increasing catches or of changes in the national data collection system.

For example, until 2009, the calculation of inland catches by Bangladesh was linked to the population increase and, as a consequence, total production grew by 67 percent between 2004 and 2009. Production reported by Myanmar has quadrupled in the last decade, increasing at an average growth rate of almost 18 percent per year, gaining 11 positions in the global ranking of major producer countries, and exceeding one million tonnes in 2010. The gathering of India's catch statistics is complex as the Ministry of Agriculture has to receive and assemble data from 28 states, which often have different systems of collecting and reporting data. It is very difficult to discern whether the dramatic growth (179 percent) in inland catches between 2004 and 2010 is ascribable to a real increase, to overestimation or to improvement in the data collection system of some of these states.

Inland water capture production in the other continents shows different trends. Uganda and the United Republic of Tanzania, fishing mostly in the African Great Lakes, and Nigeria and Egypt, with river fisheries, remain the main producers in Africa. Catches in several South American countries (e.g. Argentina, Colombia, Paraguay and Venezuela as well as in North American ones have been reported as shrinking. Increased European production between 2004 and 2010 is all attributable to a rise of almost 50 percent in catches of the Russian Federation. Inland fishery production is marginal in countries in Oceania.

More than half of the global inland water capture production is still reported as "catches unidentified by species". However, in recent years, several countries have made efforts to improve the quality of their inland catch statistics and collect data at a finer species breakdown. In the last ten years, the increase in inland water species with statistics in the FAO database has been five times that for marine species. Moreover, the percentage of inland water species in total species has improved, reaching 12.3 percent in 2010 – a value very close to the share (12.7 percent) of inland water catches in global catches in that year.

Global Aquaculture Production

Global aquaculture production has continued to grow in the new millennium, albeit more slowly than in the 1980s and 1990s. In the course of half a century or so, aquaculture has expanded from being almost negligible to fully comparable with capture production in terms of feeding people in the world. Aquaculture has also evolved in terms of technological innovation and adaptation to meet changing requirements.

World aquaculture production attained another all-time high in 2010, at 60 million tonnes (excluding aquatic plants and non-food products), with an estimated total value of US$119 billion. One-third of the world's farmed food fish harvested in 2010 was achieved without the use of feed, through the production of bivalves and filter-feeding carps. When farmed aquatic plants and non-food products are included, world aquaculture production in 2010 was 79 million tonnes, worth US$125 billion.

About 600 aquatic species are raised in captivity worldwide for production in a variety of farming systems and facilities of varying input intensities and technological sophistication, using freshwater, brackish water and marine water. Aquaculture also contributes substantially, with hatchery-produced seeds for stocking, to culture-based capture fishery production, particularly in inland waters.

However, the stage of development and the distribution of aquaculture production remain imbalanced in all regions. A few developing countries in Asia and the Pacific, sub-Saharan Africa and South America have made considerable progress in aquaculture development in recent years and they are becoming significant or major producers in their respective regions. However, the disparity remains huge across the continents and georegions, as well as among countries of comparable natural conditions in the same region, with aquaculture in many of the LDCs yet to make a significant contribution to national food and nutrition security.

In 2010, FAO recorded 181 countries and territories with aquaculture production, and 9 countries and territories not reporting production in 2010 but with production recorded previously. Of these 190 countries and territories, about 30 percent of them, including a few major producers in Asia and Europe, had failed to report any statistics on national aquaculture production even a year after the 2010 reference year. Less than 30 percent of them were able to report national data covering grow-out production broken down by culture environment and farming method or in terms of seed production and culture areas and facilities. More than 40 percent of them reported national data in varying degrees of completeness, data quality and timeliness of reporting. To compensate for such gaps, FAO made estimates using information available from additional sources where possible.

Global statistics are still lacking on: (i) non-food aquaculture production, including live bait for fishing, live ornamental species (animals and plants) and ornamental products (pearls and shells); (ii) fishes cultured as feed for certain carnivorous farmed species; (iii) culture of biomass of many species (such as plankton, *Artemia* and marine worms) for use as feed in aquaculture hatcheries and grow-out operations; (iv) aquaculture hatchery and nursery outputs for ongrowing in captivity or stocking to the wild; and (v) inputs in terms of captured wild fish ongrown in captivity. These practices are often specialized and segmented standalone operations of local importance in many countries. There is an urgent need to improve and expand national and international aquaculture statistics collection and reporting schemes in order to have a full understanding of aquaculture in accordance with the commitments made by States in 2003 in adopting the FAO Strategy and Outline Plan for Improving Information on Status and Trends of Aquaculture.

Food Fish Production

In 2010, global production of farmed food fish was 59.9 million tonnes, up by 7.5 percent from 55.7 million tonnes in 2009 (32.4 million tonnes in 2000). Farmed food fish include finfishes, crustaceans, molluscs, amphibians (frogs), aquatic reptiles (except crocodiles) and other aquatic animals (such as sea cucumbers, sea urchins, sea squirts and jellyfishes) that are indicated as fish throughout this document. The reported grow-out production from aquaculture is almost entirely destined for human consumption.

In the last three decades (1980–2010), world food fish production of aquaculture has expanded by almost 12 times, at an average annual rate of 8.8 percent. Aquaculture enjoyed high average annual growth rates of 10.8 percent and 9.5 percent in the 1980s and 1990s, respectively, but has since slowed to an annual average of 6.3 percent.

Since the mid-1990s, aquaculture has been the engine driving growth in total fish production as global capture production has levelled off. Its contribution to world total fish production climbed steadily from 20.9 percent in 1995 to 32.4 percent in 2005 and 40.3 percent in 2010. Its contribution to world food fish production for human consumption was 47 percent in 2010 compared with only 9 percent in 1980.

The growth rate in farmed food fish production from 1980 to 2010 far outpaced that for the world population (1.5 percent), resulting in average annual per capita consumption of farmed fish rising by almost seven times, from 1.1 kg in 1980 to 8.7 kg in 2010, at an average rate of 7.1 percent per year. The total farmgate value of food fish production from aquaculture is estimated at US$119.4 billion for 2010. This might be overstated considering that some countries reported values other than first-sale prices (e.g. using retail, export or processed product prices).

World aquaculture production is vulnerable to adverse impacts of natural, socio-economic, environmental and technological conditions. For example, marine cage culture of Atlantic salmon in Chile, oyster farming in Europe (notably France), and marine shrimp farming in several countries in Asia, South America and Africa have experienced high mortality caused by disease outbreaks in recent years, resulting in partial or sometimes total loss of production. Countries prone to natural disasters suffer seriously from production damage or losses caused by floods, droughts, tropical storms and, less frequently, earthquakes. Water pollution has increasingly threatened production in some newly industrialized and rapidly urbanizing areas. In 2010, aquaculture in China suffered production losses of 1.7 million tonnes (worth US$3.3 billion) caused by diseases (295 000 tonnes), natural disasters (1.2 million tonnes), pollution (123 000 tonnes), etc. Disease outbreaks virtually wiped out marine shrimp farming production in Mozambique in 2011.

Production Among Regions

Asia accounted for 89 percent of world aquaculture production by volume in 2010, up from 87.7 percent in 2000. The contribution of freshwater aquaculture has gradually increased, up to 65.6 percent in 2010 from around 60 percent during 1990s. In terms of volume, Asian aquaculture is dominated by finfishes (64.6 percent), followed by molluscs (24.2 percent), crustaceans (9.7 percent) and miscellaneous species (1.5 percent). The share of non-fed species farmed in Asia was 35 percent (18.6 million tonnes) in 2010 (compared with 50 percent in 1980). The contribution of China to world aquaculture production volume in 2010 declined to 61.4 percent from its highest level of about 66 percent in the period 1996–2000. Other major producers in Asia (India, Viet Nam, Indonesia, Bangladesh, Thailand, Myanmar, the Philippines and Japan) are among the world's top producers.

In the Americas, the share of freshwater aquaculture in total production declined from 54.8 percent in 1990 to 37.9 percent in 2010. In North America, aquaculture has ceased expanding in recent years, but in South America it has shown strong and continuous growth, particularly in Brazil and Peru. In terms of volume, aquaculture in North and South America is dominated by finfishes (57.9 percent), crustaceans (21.7 percent) and molluscs (20.4 percent). Bivalve production fluctuated between 14 and 21 percent of total aquaculture production in the 1990s and 2000s, after dropping rapidly in the 1980s from 48.5 percent.

In Europe, the share of production from brackish and marine waters increased from 55.6 percent in 1990 to 81.5 percent in 2010, driven by marine cage culture of Atlantic salmon and other species. Several important producers in Europe have recently ceased expanding or have even contracted, particularly in the marine bivalve sector. In 2010, finfishes accounted for three-quarters of all European aquaculture production, and molluscs one-quarter. The share of bivalves in total production decreased continuously from 61 percent in 1980 to 26.2 percent in 2010.

Africa has increased its contribution to global production from 1.2 percent to 2.2 percent in the past ten years, albeit from a very low base. The share of freshwater aquaculture in the region fell from 55.2 percent to 21.8 percent in the 1990s, largely reflecting the strong growth in brackish-water culture in Egypt, but it recovered in the 2000s, reaching 39.5 percent in 2010 as a result of rapid development in freshwater fish farming in sub-Saharan Africa, most notably in Nigeria, Uganda, Zambia, Ghana and kenya. African aquaculture production is overwhelmingly dominated by finfishes (99.3 percent by volume), with only a small fraction from marine shrimps (0.5 percent) and marine molluscs (0.2 percent). In spite of some limited successes, the potential for bivalve production in marine waters remains almost completely unexplored.

Oceania is of relatively marginal importance in global aquaculture production. Production from this region consists mainly of marine molluscs (63.5 percent) and finfishes (31.9 percent), while crustaceans (3.7 percent, mostly marine shrimps) and other species (0.9 percent) constitute less than 5 percent of its total production. Marine bivalves accounted for about 95 percent of the total produced in the first half of 1980s but, reflecting the development of the finfish culture sector (especially Atlantic salmon in Australia and chinook salmon in New Zealand), they currently account for less than 65 percent of the region's total production. Freshwater aquaculture accounts for less than 5 percent of the region's production.

The global distribution of aquaculture production across the regions and countries of different economic development levels remains imbalanced. In 2010, the top ten producing countries accounted for 87.6 percent by quantity and 81.9 percent by value of the world's farmed food fish. At the regional level, production is also concentrated in a few major producers.

The LDCs, mostly in sub-Saharan Africa and in Asia, and home to 20 percent of the world's population (1.4 billion people), remain very small in terms of their share of world aquaculture production (4.1 percent by quantity and 3.6 percent by value). The major producers in the LDCs in 2010 include Bangladesh, Myanmar, Uganda, the Lao People's Democratic Republic (82 100 tonnes), Cambodia (60 000 tonnes) and Nepal (28 200 tonnes).

While aquaculture production has shown strong growth in developing countries, particularly in Asia, annual growth rates in developed industrialized countries averaged only 2.1 percent and 1.5 percent in the 1990s and 2000s, respectively. In 2010, they produced collectively 6.9 percent (4.1 million tonnes) by quantity and 14 percent (US$16.6 billion) by value of world farmed food fish production, compared with 21.9 percent and 32.4 percent in 1990. Aquaculture production has contracted or stagnated in Japan, the United States of America, Spain, France, the United kingdom of Great Britain and Northern Ireland, Canada and Italy. An exception is Norway, where, thanks to the farming of Atlantic salmon in marine cages, aquaculture production grew from 151 000 tonnes in 1990 to more than one million tonnes in 2010, at an average growth rate of 12.6 percent in the 1990s and 7.5 percent in the 2000s.

In the recent past, some developing countries in Asia and the Pacific (Myanmar and Papua New Guinea), sub-Saharan Africa (Nigeria, Uganda, kenya, Zambia and Ghana) and South America (Ecuador, Peru and Brazil) have made rapid progress to become significant or major aquaculture producers in their regions.

Immediately after their independence more than two decades ago, countries in the former Soviet Union were producing an annual total of almost 350 000 tonnes of food fish from aquaculture. However, production capacity in all these countries deteriorated rapidly in the 1990s to about one-third of its original level. In spite of starting to recover in the 2000s, their combined total production in 2010 amounted to only 59 percent of that in 1988. The lost capacity, especially in hatchery and nursery output, has also had a negative impact on inland culture-based capture fisheries. While Armenia, Belarus, Estonia and Republic of Moldova have exceeded their 1988 production levels, and output in Lithuania and the Russian Federation is at more than 80 percent of its original 1998 level, other countries remain at one-third or less of their 1988 production levels. In 2010, farmed fish production in kazakhstan and Turkmenistan was less than 5 percent of that before independence.

Production With and Without Feed

While feed is generally perceived to be a major constraint to aquaculture development, one-third of all farmed food fish production, 20 million tonnes, is currently achieved without artificial feeding. Oysters, mussels, clams, scallops and other bivalve species are grown with food materials that occur naturally in their culture environment in the sea and lagoons. Silver carp and bighead carp feed on planktons proliferated through intentional fertilization and the wastes and leftover feed materials of fed species grown in the same multispecies polyculture systems. Rice–fish farming has long been a common practice, particularly in Asia.

However, the percentage of non-fed species in world production has declined gradually from more than 50 percent in 1980 to the present level of 33.3 percent, strongly dominated by changing practices in Asia. This reflects the relatively faster growth in the fed-species culture subsector supported by, among others, the development and improved availability of formulated aquaculture feeds for finfishes and crustaceans.

Some fed species grow on a mixture of natural food proliferated from fertilization and supplementary feeds. If the non-fed portion in their total production were considered, the non-fed portion of world production of all farmed food fish would be higher than the aforesaid 33.3 percent. Owing to the unavailability of information and data needed for the calculation, the said percentage does not include: (i) the non-fed portion of production of some fed species (such as milkfish that grow partially on algal aggregates known as "lab-lab" proliferated through fertilization in culture ponds); and (ii) the non-fed filter feeding carps reported by some producers in aggregation with other species and treated wholly as fed species.

In terms of food security, producers in Asia, especially China, Viet Nam, India, Indonesia and Bangladesh, have benefited from the development of culture of low-trophic-level species, such as carps and barbs, tilapias and *Pangasius* catfish, in easing dependence on high-protein feeds, and thus reduced the vulnerability of their sectors to externalities. Grass carp, the world's most-produced finfish species from aquaculture, is grown partially with cultivated and wild-collected "pastures", instead of using formulated feeds only.

The production of 253 000 tonnes of highly carnivorous Mandarin fish (*Siniperca chuatsi*), which feeds on live prey only, was achieved by feeding them with low-trophic-level carp fingerlings grown with low-protein feeds plus pond fertilization.

Comparable in quantity with the total production of farmed rainbow trout in Europe (257 200 tonnes), or the combined world production of gilthead seabream and European seabass (265 100 tonnes), Mandarin fish production has been assumed to be dependent on fishmeal and fish oil for feed, and this now needs reconsideration. As discussed above, part of its production could be treated as the non-fed portion of fed species production. In sub-Saharan Africa, the carnivorous North African catfish (*Clarias gariepinus*) has replaced tilapia as the most-produced fish in aquaculture since 2004. The progressive dominance of catfish species in aquaculture is particularly pronounced in Nigeria and Uganda. Being the largest producer of catfish in Africa, Nigeria even imports catfish feeds from as far away as Northern Europe.

Production by Culture Environment

Aquaculture production uses freshwater, brackish water and full-strength marine water as culture media. Data available at FAO show that, in terms of quantity, the percentage of production from freshwater rose from less than 50 percent before the 1980s to almost 62 percent in 2010, with the share of marine aquaculture production declining from more than 40 percent to just above 30 percent. In 2010, freshwater aquaculture was the source of 58.1 percent of global production by value. Brackish-water aquaculture yielded only 7.9 percent of world production in terms of quantity but accounted for 12.8 percent of total value because of the relatively high-valued marine shrimps cultured in brackish-water ponds. Marine water aquaculture accounted for about 29.2 percent of world aquaculture production by value.

The average annual growth rate for freshwater aquaculture production from 2000 to 2010 was 7.2 percent, compared with 4.4 percent for marine aquaculture production. Freshwater fish farming has been a relatively easy entry point for practising aquaculture in developing countries, particularly for small-scale producers.As such, freshwater aquaculture is expected to contribute further to total aquaculture production in the 2010s.

The share of brackish-water aquaculture production has been stable, ranging between 6 and 8 percent, for most of the time. An exception was in the 1980s and early 1990s when accelerated development of brackish-water culture of marine shrimp species, particularly in coastal regions of Asia and South America, led to brackish-water aquaculture reaching 8–10 percent of total production. However, in the period 1994– 2000, world marine shrimp farming was hit by disease outbreaks in Asia and South America, and the share of brackish-water production fell to 6 percent.

At the global level, the composition and types of farmed species differ greatly among the three culture environments, and they have also undergone changes within environments over the years.

Freshwater aquaculture production (36.9 million tonnes) was overwhelmingly dominated by finfishes (91.7 percent, 33.9 million tonnes) in 2010, as in the past. Crustaceans accounted for 6.4 percent, and all other types of species contributed only 1.9 percent. The development of freshwater farming of crustaceans and other species (such as soft-shell turtles and frogs) in the past two decades has slightly eroded the dominance of finfish in production. The share of diadromous fishes, including rainbow trout and other salmonids, eels and sturgeons, shrank from 6.3 percent in 1990 to 2.5 percent in 2010.

Brackish-water aquaculture production (4.7 million tonnes) consisted of crustaceans (57.2 percent, 2.7 million tonnes), freshwater fishes (18.7 percent), diadromous fishes (15.4 percent), marine fishes (6.5 percent) and marine molluscs (2.1 percent) in 2010. More the 99 percent of the crustaceans were marine shrimps. The share of freshwater fishes has increased dramatically in the past two decades, driven largely by rapid development in Nile tilapia and other species in Egypt. Milkfish and barramundi remain important but their combined share has dropped significantly. Salmonids and eels are also cultured in brackish-water in small quantities.

Marine-water aquaculture production (18.3 million tonnes) consists of marine molluscs (75.5 percent, 13.9 million tonnes), finfishes (18.7 percent, 3.4 million tonnes), marine crustaceans (3.8 percent) and other aquatic animals (2.1 percent), e.g. sea cucumbers, and sea urchins. The share of molluscs (mostly bivalves, e.g. oysters, mussels, clams, cockles, arkshells and scallops) declined from 84.6 percent in 1990 to 75.5 percent in 2010, reflecting the rapid growth in finfish culture in marine water, which grew at an average annual rate of 9.3 percent from 1990 to 2010 (seven times faster than the rate for molluscs). Salmonid production, particularly Atlantic salmon, increased dramatically from 299 000 tonnes in 1990 to 1.9 million tonnes in 2010, at an average annual rate exceeding 9.5 percent. Other finfish species also increased rapidly, from 278 000 tonnes in 1990 to 1.5 million tonnes in 2010, at an average annual rate exceeding 8.6 percent. Other finfish species cultured in marine water include amberjacks, seabreams, seabasses, croakers, grouper, drums, mullets, turbot and other flatfishes, snappers, cobia, pompano, cods, puffers and tunas.

Species Produced in Aquaculture

In 2010, the composition of world aquaculture production was: freshwater fishes (56.4 percent, 33.7 million tonnes), molluscs (23.6 percent, 14.2 million tonnes), crustaceans (9.6 percent, 5.7 million tonnes), diadromous fishes (6.0 percent, 3.6 million tonnes), marine fishes (3.1 percent, 1.8 million tonnes) and other aquatic animals (1.4 percent, 814 300 tonnes).

Production of freshwater fishes has always been dominated by carps (71.9 percent, 24.2 million tonnes, in 2010). Among carps, 27.7 percent are non-fed filter-feeders and the rest are fed with low-protein feeds. Production of tilapias has a wide distribution, and 72 percent are raised in Asia (particularly in China and Southeast Asia), 19 percent in Africa, and 9 percent in America. Viet Nam dominates production of omnivorous *Pangasius* catfishes although there are other producers, such as Indonesia and Bangladesh. World production of *Pangasius* catfish may be understated because booming production in India has yet to be reflected in statistics. In 2010, Asia accounted for 73.7 percent of the production of other catfish species, America took its share to 13.5 percent (with channel catfish production), leaving 12.3 percent of production in Africa (dominated by North African catfish). Carnivorous species such as perches, basses and snakeheads accounted for only 2.6 percent of all freshwater fish produced in 2010.

Since the beginning of 1990s, more than half of the world production of diadromous fishes has come from salmonids, and the share peaked at 70.4 percent in 2001 before declining slightly in the face of increased milkfish production in Asia. The production of Japanese and European eels, mostly raised in East Asia and to a much lesser extent in Europe, has remained at about 270 000 tonnes in recent years. Limited by the supply of seeds, the chances of a significant increase in coming years appear remote. Other eel species have been tested with wild-collected seeds with only limited success. Culture of sturgeons, for meat and for caviar, has risen steadily in Asia, Europe and America although production is still small. An increased number of farming systems with sophisticated equipment requiring high investment have been set up to target caviar production in some countries.

World production of marine fishes is more evenly distributed across the cultured species. However, almost half a million tonnes, or one-quarter of global production, are reported without identifying the species, particularly by a few top producers from Asia. There is evidence that production of European seabass and gilthead seabream has been significantly under-reported in some areas in the Mediterranean.

World aquaculture production of crustaceans in 2010 consisted of freshwater species (29.4 percent) and marine species (70.6 percent). The production of marine species is dominated by white leg shrimp (*Penaeus vannamei*), including substantial production in freshwater. In sharp contrast, the giant tiger prawn has lost importance in the last decade. Major freshwater species include red swamp crayfish, Chinese mitten crab, oriental shrimp and giant river prawn.

Regarding molluscs, aquaculture production of clams and cockles has increased much faster than that of other species groups. In 1990, clam and cockle production was half that of oysters, but by 2008 it exceeded oysters and became the most-produced species group of molluscs. Among other aquatic animals, production of sea cucumbers and soft-shell turtles has increased rapidly.

Use of Aquatic species in aquaculture production

The number of species recorded in FAO aquaculture production statistics increased to 541 species and species groups in 2010, including 327 finfishes (5 hybrids), 102 molluscs, 62 crustaceans, 6 amphibians and reptiles, 9 aquatic invertebrates and 35 algae. The increase reflects improvements in data collection and reporting at the international and national levels, as well as the farming of new species, including hybrids. In view of the high degree of species aggregation reported by many countries, it is estimated that aquaculture production worldwide uses about 600 aquatic food fish and algae species.

Exotic aquatic species have been widely introduced and used for mass production in aquaculture, and their use is particularly common and important in Asian countries. Successful internationally introduced species for finfishes include tilapias from Africa (especially Nile tilapia), Chinese carps (silver carp, bighead carp and grass carp), Atlantic salmon (*Salmo salar*), *Pangasius* catfishes (*Pangasius* spp.), largemouth black bass (*Micropterus salmoides*), turbot (*Scophthalmus maximus*), piarapatinga (*Piaractus brachypomus*), pacu (*Piaractus mesopotamicus*), and rainbow trout (*Oncorhynchus mykiss*).

Measured by production, white leg shrimp is the most successful internationally introduced marine crustacean species for aquaculture. In 2010, it accounted for 71.8 percent of world production of all farmed marine shrimp species, of which 77.8 percent was produced in Asia (with the rest in its native home in America). Some shrimp-farming countries maintain bans on the farming of this exotic species, and Bangladeshi shrimp growers and seafood exporters have recently requested a lifting of the ban. Red swamp crayfish (*Procambarus clarkii*) from North America and giant river prawn (*Macrobrachium rosenbergii*) from South and Southeast Asia have also become important for freshwater culture in countries foreign to these species.

A significant part of the global production of marine molluscs, particularly in Europe and America, relies on the widely introduced Japanese carpet shell (*Ruditapes philippinarum*, also known as Manila clam) and Pacific cupped oyster (*Crassostrea gigas*). China now produces large quantities of Atlantic bay scallop (*Argopecten irradians*) and Yesso scallop (*Patinopecten yessoensis*).

A considerable number of hybrids, most notably of finfish, are used in aquaculture, especially in countries with a relatively high level of development in aquaculture technologies. Commercially farmed hybrids include: sturgeons (such as beluga *Huso huso* x starlet sturgeon *Acipenser ruthenus* known as "bester") in Asia and Europe;*Carassius* spp., snakeheads and groupers in China; characins in South America; and freshwater catfishes

(*Clarias gariepinus* x *Heterobrachus longifilis*) in Africa and Europe. The culture of hybrid tilapias is particularly common around the world. The hybrid of *Oreochrom aureus* x *O. niloticus* (with a high percentage of male offspring) is farmed in China, and the saline-resistant hybrid of *O. niloticus* x *O. mossambicus* in the Philippines.

Five finfish hybrids have been recorded with national production statistics and FAO estimates, indicating world production levels in 2010 of 333 300 tonnes of blue and Nile tilapia hybrid (*Oreochrom aureus* x *O. niloticus*, in China and in Panama), 116 900 tonnes of *Clarias* catfish hybrid (*Clarias gariepinus* x *C. macrocephalus*, in Thailand), 21 600 tonnes of "tambacu" hybrid (*Piaractus mesopotamicus* x *Colossoma macropomum*, in Brazil), 4 900 tonnes of "tambatinga" hybrid (*Colossoma macropomum* x *Piaractus brachypomus*, in Brazil) and 4 200 tonnes of striped bass hybrid (*Morone chrysops* x *M. saxatilis*, in the United States of America, Italy and Israel).

Aquatic Plant (algae) Production

To date, only aquatic algae have been recorded globally in farmed aquatic plant production statistics. Global production has been dominated by marine macroalgae, or seaweeds, grown in both marine and brackish waters.

Aquatic algae production by volume increased at average annual rates of 9.5 percent in the 1990s and 7.4 percent in the 2000s – comparable with rates for farmed aquatic animals – with production increasing from 3.8 million tonnes in 1990 to 19 million tonnes in 2010. Cultivation has overshadowed production of algae collected from the wild, which accounted for only 4.5 percent of total algae production in 2010.

Following downward adjustments by FAO of the estimated value of several major species from a few major producers with incomplete reported data, the estimated total value of farmed algae worldwide has been reduced for a number of years in the time series. The total value of farmed aquatic algae in 2010 is estimated at US$5.7 billion, while that for 2008 is now re-estimated at US$4.4 billion.

A few species dominate algae culture, with 98.9 percent of world production in 2010 coming from Japanese kelp (*Saccharina/Laminaria japonica*) (mainly in the coastal waters of China), *Eucheuma* seaweeds (a mixture of *Kappaphycus alvarezii*, formerly known as *Eucheuma cottonii*, and *Eucheuma* spp.), *Gracilaria* spp., nori/laver (*Porphyra spp.*), wakame

(*Undaria pinnatifida*) and unidentified marine macroalgae species (3.1 million tonnes, mostly from China). The remainder consists of marine macroalgae species farmed in small quantities (such as *Fusiform sargassum* and *Caulerpa* spp.) and microalgae cultivated in freshwater (mostly *Spirulina* spp., plus a small fraction of *Haematococcus pluvialis*). The production increase is most obvious in the farming of *Eucheuma* seaweeds.

In sharp contrast to fish aquaculture, the cultivation of aquatic algae is practised in far fewer countries. Only 31 countries and territories are recorded with algae farming production in 2010, and 99.6 percent of global cultivated algae production comes from just eight countries: China (58.4 percent, 11.1 million tonnes), Indonesia (20.6 percent, 3.9 million tonnes), the Philippines (9.5 percent, 1.8 million tonnes), the Republic of korea (4.7 percent, 901 700 tonnes), Democratic People's Republic of korea (2.3 percent, 444 300 tonnes), Japan (2.3 percent, 432 800 tonnes), Malaysia (1.1 percent, 207 900 tonnes) and the United Republic of Tanzania (0.7 percent, 132 000 tonnes).

Status of Fishery Resources

Marine Fisheries

The world's marine fisheries have experienced different stages, increasing from 16.8 million tonnes in 1950 to a peak of 86.4 million tonnes in 1996, and then declining to stabilize at about 80 million tonnes, with interannual fluctuations. Global recorded production was 77.4 million tonnes in 2010. Of the marine areas, the Northwest Pacific had the highest production with 20.9 million tonnes (27 percent of the global marine catch) in 2010, followed by the Western Central Pacific with 11.7 million tonnes (15 percent), the Northeast Atlantic with 8.7 million tonnes (11 percent), and the Southeast Pacific, with a total catch of 7.8 million tonnes (10 percent).

The proportion of non-fully exploited stocks has decreased gradually since 1974 when the first FAO assessment was completed. In contrast, the percentage of overexploited stocks increased, especially in the late 1970s and 1980s, from 10 percent in 1974 to 26 percent in 1989. After 1990, the number of overexploited stocks continued to increase, albeit at a slower rate. The fraction of fully exploited stocks demonstrates the smallest change over time. Its percentage was stable at about 50 percent from 1974 to 1985, then dropped to 43 percent in 1989 before gradually increasing to 57.4 percent in 2009.

By definition, the fully exploited stocks produce catches that are at or very close to their maximum sustainable production. Therefore, they have no room for further expansion in catch, and may even be at some risk of decline unless properly managed. Among the remaining stocks, 29.9 percent were overexploited, and 12.7 percent non-fully exploited in 2009. Overexploited stocks produce lower yields than their biological and ecological potential. They require strict management plans to rebuild stock abundance and restore full and sustainable productivity. The Johannesburg Plan of Implementation that resulted from the World Summit on Sustainable Development demands that all these stocks be restored to the level that can produce maximum sustainable yield by 2015. The non-fully exploited stocks are under relatively low fishing pressure and have some potential to increase their production. However, these stocks often do not have a high production potential. The potential for increase in catch may be generally limited. Nevertheless, proper management plans should be established before increasing the exploitation rate of these non-fully exploited stocks in order to avoid following the same track of overfishing as many currently overexploited stocks.

Most of the stocks of the top ten species, which account in total for about 30 percent of the world marine capture fisheries production, are fully exploited and, therefore, have no potential for increases in production, while some stocks are overexploited and increases in their production may be possible if effective rebuilding plans are put in place. The two main stocks of anchoveta in the Southeast Pacific, Alaska pollock (*Theragra chalcogramma*) in the North Pacific and blue whiting (*Micromesistius poutassou*) in the Atlantic are fully exploited. Atlantic herring (*Clupea harengus*) stocks are fully exploited in both the Northeast and Northwest Atlantic. Japanese anchovy (*Engraulis japonicus*) in the Northwest Pacific and Chilean jack mackerel (*Trachurus murphyi*) in the Southeast Pacific are considered to be overexploited. Chub mackerel (*Scomber japonicus*) stocks are fully exploited in the Eastern Pacific and the Northwest Pacific. The largehead hairtail (*Trichiurus lepturus*) was estimated in 2009 to be overexploited in the main fishing area in the Northwest Pacific.

The total catch of tuna and tuna-like species was about 6.6 million tonnes in 2010. The principal market tuna species – albacore, bigeye, bluefin (three species), skipjack and yellowfin – contributed 4.3 million tonnes, maintaining approximately the same level since 2002. About 70 percent of these catches were from the Pacific. The skipjack was the most productive principal market tuna, contributing about 58 percent, and the yellowfin and bigeye were the other two productive species, contributing about 27 and 8 percent, respectively, to the 2010 catch of principal tunas. Bigeye, Atlantic bluefin, Pacific bluefin, southern bluefin and yellowfin tunas have all shown a gradual decline in catch after reaching historical peaks.

Among the seven principal tuna species, one-third were estimated to be overexploited, 37.5 percent were fully exploited, and 29 percent non-fully exploited in 2009. Although skipjack tuna continued its increasing trend up to 2009, further expansion should be closely monitored, as it may negatively affect bigeye and yellowfin tunas (multispecies fisheries). Only for very few stocks of the principal tuna species is their status unknown or very poorly known. In the long term, the status of tuna stocks (and consequently catches) may further deteriorate unless there are significant improvements in their management. This is because of the substantial demand for tuna and the significant overcapacity of tuna fishing fleets.

World marine fisheries have gone through significant changes since the 1950s. Accordingly, the exploitation level of fish resources and their landings have also varied over time. The temporal pattern of landings differs from area to area depending on the level of urban development and changes that countries surrounding that area have experienced. In general, they can be divided into three groups, i.e. one characterized by oscillations in the catches, another by an overall declining trend following historical peaks, and a third with increasing catch trends.

The first group includes those FAO areas that have demonstrated oscillations in total catch, i.e. the Eastern Central Atlantic (Area 34), Northeast Pacific (Area 67), Eastern Central Pacific (Area 77), Southwest Atlantic (Area 41), Southeast Pacific (Area 87), and Northwest Pacific (Area 61). These areas have provided about 52 percent of the world's total marine catch on average in the last five years. Several of these areas include upwelling regions that are characterized by high natural variability.

The second group consists of areas that have demonstrated a decreasing trend in catch since reaching a peak at some time in the past. This group has contributed 20 percent of global marine catch on average in the last five years, and includes the Northeast Atlantic (Area 27), Northwest Atlantic (Area 21), Western Central Atlantic (Area 31), Mediterranean and Black Sea (Area 37), Southwest Pacific (Area 81), and Southeast Atlantic (Area

47). It should be noted that lower catches in some cases reflect fisheries management measures that are precautionary or aim at rebuilding stocks, and this situation should, therefore, not necessarily be interpreted as negative.

The third group comprises the FAO areas that have shown continuously increasing trends in catch since 1950. There are only three areas in this group: Western Central Pacific (Area 71), Eastern (Area 57) and Western Indian Ocean (Area 51). They have contributed 28 percent of the total marine catch on average over the last five years. However, in some regions, there is still high uncertainty about the actual catches owing to the poor quality of statistical reporting systems in coastal countries.

The Northwest Pacific has the highest production among the FAO statistical areas. Its total catch fluctuated between about 17 and 24 million tonnes in the 1980s and 1990s, and was about 21 million tonnes in 2010. Small pelagics are the most abundant category in this area, with the Japanese anchovy providing 1.9 million tonnes in 2003 but having since declined to about 1.1 million tonnes in 2009 and 2010. Other important contributors to the total catch in the area are the largehead hairtail, considered overexploited, and the Alaska pollock and chub mackerel, both considered fully exploited. Squids, cuttlefish and octopuses are important species, yielding 1.3 million tonnes in 2010.

The Eastern Central Pacific has shown a typical oscillating pattern in its total catch since 1980 and produced about 2 million tonnes in 2010. The Southeast Pacific has had a large interannual variation with a generally declining trend since 1993. There have been no major changes in the state of exploitation of stocks in these two areas, which are characterized by a large proportion of small pelagic species and great fluctuations in catches. The most abundant species in the Southeast Pacific are the anchoveta, the Chilean jack mackerel and the South American pilchard or sardine (*Sardinops sagax*), accounting for more than 80 percent of the current and historical catches, while in the Eastern Central Pacific the most abundant species are California pilchard and Pacific anchoveta. A moderate El Niño developed in 2009 and continued throughout the equatorial Pacific in the first few months of 2010. Deep tropical convection remained enhanced across central and eastern parts of the tropical Pacific with relatively mild impacts reported on the state of stocks and fisheries in the eastern Pacific.

For the Eastern Central Atlantic, total catches, which have fluctuated since the 1970s, were about 4 million tonnes in 2010, about the same as the 2001 peak. The small pelagic species constitute almost 50 percent of the landings, followed by "miscellaneous coastal fishes". The single most important species in terms of landings is sardine (*Sardina pilchardus*) with landings in the range of 600 000–900 000 tonnes in the last ten years. The sardine in Zone C (Cape Bojador and southwards to Senegal) is still considered non-fully exploited; otherwise, most of the pelagic stocks are considered fully exploited or overexploited, such as the sardinella stocks in Northwest Africa and in the Gulf of Guinea. The demersal fish resources are to a large extent fully exploited to overexploited in most of the area, and the white grouper stock (*Epinephelus aenus*) in Senegal and Mauritania remains in a severe condition. The status of some of the deepwater shrimp stocks seems to have improved and they are now considered fully exploited, whereas the other shrimp stocks in the region range between fully exploited and overexploited. The commercially important octopus (*Octopus vulgaris*) and cuttlefish (*Sepia* spp.) stocks remain overexploited. Overall, the Eastern Central Atlantic has 43 percent of its assessed stocks fully exploited, 53 percent overexploited and 4 percent non-fully exploited, a situation warranting attention for improvement in management.

In the Southwest Atlantic, total catches have fluctuated around 2 million tonnes after a period of increasing catches ended in the mid-1980s. Major species such as Argentina hake and Brazilian sardinella are still estimated to be overexploited, although there seem to be some signs of recovery for the latter. The catch of Argentina shortfin squid was only one-fourth of its peak level in 2009 and considered fully exploited to overexploited. In this area, 50 percent of the monitored fish stocks were overexploited, 41 percent fully exploited and the remaining 9 percent considered non-fully exploited.

The Northeast Pacific produced 2.4 million tonnes of fish in 2010, similar to the production level in the early 1970s, although more than 3 million tonnes was seen in the late 1980s. Cods, hakes and haddocks are the largest contributors to its catch.

In this area, only 10 percent of fish stocks were estimated to be overexploited, with 80 percent fully exploited, and another 10 percent non-fully exploited.

In the Northeast Atlantic, total catch appeared to have a decreasing trend after 1975, with a recovery in the 1990s, and was 8.7 million tonnes in 2010. The blue whiting stock decreased rapidly from the peak of 2.4

million tonnes in 2004 to only 0.6 million tonnes in 2009. Fishing mortality has been reduced in cod, sole and plaice, with recovery plans in place for the major stocks of these species. The Arctic cod spawning stock was particularly large in 2008, having recovered from the low levels observed in the 1960s–1980s. Similarly, the Arctic saithe and haddock stocks have increased to high levels, although stocks elsewhere remain fully exploited or overexploited. The largest sand eel and capelin stocks remain overexploited. Concern remains for redfishes and deep-water species for which data are limited and which are likely to be vulnerable to overfishing. Northern shrimp and Norway lobster are generally in good condition, but there are indications that some stocks are being overexploited. Recently, maximum sustainable yield has been adopted as the standard basis for reference points. Overall, 62 percent of assessed stocks are fully exploited, 31 percent overexploited, and 7 percent non-fully exploited.

Although fishery resources in the Northwest Atlantic continue to be under stress from previous and/or current exploitation, some stocks have recently shown signs of renewal in response to an improved management regime in the last decade (e.g. Greenland halibut, yellowtail flounder, Atlantic halibut, haddock, spiny dogfish). However, some historical fisheries such as cod, witch flounder and redfish still evidence lack of recovery, or limited recovery, which may be the result of unfavourable oceanographic conditions and the high natural morality caused by increasing numbers of seals, mackerel and herring. These factors appear to have affected fish growth, reproduction and survival. Conversely, invertebrates remain at near record levels of abundance. The Northwest Atlantic has 77 percent of stocks fully exploited, 17 percent overexploited and 6 percent non-fully exploited.

The Southeast Atlantic is a typical example of the group of areas that has demonstrated a generally decreasing trend in catches since the early 1970s. This area produced 3.3 million tonnes in the late 1970, but only 1.2 million tonnes were recorded in 2009. The important hake resources remain fully exploited to overexploited although there are signs of some recovery in the deepwater hake stock (*Merluccius paradoxus*) off South Africa and of the shallow-water Cape hake (*Merluccius capensis)* off Namibia, as a consequence of good recruitment years and of the strict management measures introduced since 2006. A significant change concerns the Southern African pilchard, which was at a very high biomass and estimated to be fully exploited in 2004, but which now, under unfavourable environmental conditions, has declined considerably in abundance and is now fully exploited or overexploited. In contrast, Southern African anchovy has continued to improve and its status was estimated to be fully exploited in 2009. Whitehead's round herring has not been fully exploited. The condition of Cunene horse mackerel has deteriorated, particularly off Namibia and Angola, and it was overexploited in 2009. The condition of the perlemoen abalone stock continues to be worrying, exploited heavily by illegal fishing, and it is currently overexploited and probably depleted.

The Mediterranean has maintained an overall stable catch in a difficult situation in recent years. All hake (*Merluccius merluccius*) and red mullet (*Mullus barbatus*) stocks are considered overexploited, as are probably also the main stocks of sole and most seabreams. The main stocks of small pelagic fish (sardine and anchovy) are assessed as either fully exploited or overexploited. A newly identified threat is the increasing penetration of exotic Red Sea species, which in some cases seem to be replacing native species, especially in the Eastern Mediterranean. In the Black Sea, the situation of small pelagic fish (mainly sprat and anchovy) has recovered somewhat from the drastic decline suffered in the 1990s, probably as a consequence of unfavourable oceanographic conditions, but they are still considered fully exploited to overexploited, an assessment shared with turbot, while most other stocks are probably fully exploited to overexploited. In general, the Mediterranean and Black Sea had 33 percent of assessed stocks fully exploited, 50 percent overexploited, and the remaining 17 percent non-fully exploited in 2009.

Total production in the Western Central Pacific grew continuously to a maximum of 11.7 million tonnes in 2010. This area contributes about 14 percent of the global marine production. Despite this catch trend, there are reasons for concern as regards the state of the resources, with most stocks being either fully exploited or overexploited, particularly in the western part of the South China Sea. The high catches have probably been maintained through expansion of the fisheries to new areas and possible double counting in the transshipment of catches between fishing areas, which leads to bias in estimates of production, potentially masking negative trends in stock status.

The Eastern Indian Ocean (Fishing Area 57) is still experiencing a high growth rate in catches, with a 17 percent increase from 2007 to 2010, and now totalling 7 million tonnes. The Bay of Bengal and Andaman Sea regions have seen total catches increase steadily and there are no signs of the catch levelling off. However, a very high percentage (about 42 percent) of the catches in this area are attributed to the category "marine fishes not identified", which is a cause of concern as regards the need for monitoring stock status and trends. Increased catches may in fact be due to the expansion of fishing to new areas or species. Declining catches in the fisheries within Australia's EEZ can be partly explained by a reduction in effort and in catches following a structural adjustment and a ministerial direction in 2005 aimed at ceasing overfishing and allowing overfished stocks to rebuild. The economics of fishing in this area are expected to improve in the medium and long term, but higher profits can also be expected for individual fishers in the short term because fewer vessels are operating.

In the Western Indian Ocean, total landings reached a peak of 4.5 million tonnes in 2006, but have declined slightly since, and 4.3 million tonnes were reported in 2010. A recent assessment has shown that narrow-barred Spanish mackerel (*Scomberomerus commerson*), a migratory species found in the Red Sea, Arabian Sea, Gulf of Oman, Persian Gulf, and off the coast along Pakistan and India, is overexploited. Catch data in this area are often not detailed enough for stock assessment purposes. However, the Southwest Indian Ocean Fisheries Commission conducted stock assessments for 140 species in its mandatory area in 2010 based on best-available data and information. Overall, 65 percent of fish stocks were estimated to be fully exploited, 29 percent overexploited, and 6 percent non-fully exploited in 2009.

The declining global catch over the last few years together with the increased percentage of overexploited fish stocks and the decreased proportion of non-fully exploited species around the world convey a strong message – the state of world marine fisheries is worsening and has had a negative impact on fishery production. Overexploitation not only causes negative ecological consequences, but it also reduces fish production, which further leads to negative social and economic consequences. To increase the contribution of marine fisheries to the food security, economies and well-being of the coastal communities, effective management plans must be put in place to rebuild overexploited stocks. The situation seems more critical for some highly migratory, straddling and other fishery resources that are exploited solely or partially in the high seas. The United Nations Fish Stocks Agreement that entered into force in 2001 should be used as a legal basis for management measures of the high seas fisheries.

In spite of the worrisome global situation of marine capture fisheries, good progress is being made in reducing exploitation rates and restoring overexploited fish stocks and marine ecosystems through effective management actions in some areas. In the United States of America, the Magnuson–Stevens Act and subsequent amendments have created a mandate to put overfished stocks into restoration; 67 percent of all stocks are now being sustainably harvested, while only 17 percent are still being overexploited. In New Zealand, 69 percent of stocks are above management targets, reflecting mandatory rebuilding plans for all fisheries that are still below target thresholds. Similarly, Australia reports overfishing for only 12 percent of stocks in 2009. Since the 1990s, the Newfoundland–Labrador Shelf, the Northeast United States Shelf, the Southern Australian Shelf, and California Current ecosystems have shown substantial declines in fishing pressure such that they are now at or below the modelled exploitation rate that gives the multispecies maximum sustainable yield of the ecosystem. It is critically important to understand the key elements of these and other successes and apply them well to other fisheries.

Inland Fisheries

The difficulty in assessing the state of inland capture fisheries has been noted in past editions of The State of World Fisheries and Aquaculture as well as by those working on the active management and development of inland fishery resources. Reasons for the lack of adequate assessments include: the diffuse nature of the sector, with numerous landing sites and methods of fishing;

- the large number of people involved and the seasonality of fishing effort;
- the subsistence nature of many small-scale inland fisheries;
- the fact that catch is often consumed or traded locally without entering the formal market chain;
- a lack of capacity and resources to collect adequate data;
- activities not associated with inland fishing can greatly influence the abundance of inland fishery resources, e.g. stocking from aquaculture, water diversion for agriculture and hydroelectric development.

The informative and widely cited data summarizing the state of the major marine fish stocks are virtually impossible to duplicate for the state of the world's inland fisheries. The primary reason for this is that whereas exploitation rate is the main driver affecting the state of the major marine stocks that comprise the figure, other drivers affect the status of inland fishery resources to a much greater extent. Drivers associated with habitat quantity and quality, including aquaculture in the form of stocking and competition for freshwater, influence the state of the majority of inland fishery resources much more than exploitation rates do. Water abstraction and diversion, hydroelectric development, draining wetlands, and siltation and erosion from land-use patterns can negatively affect inland fishery resources regardless of the rate of exploitation. Conversely, stock enhancement from aquaculture facilities, which is widely practised in inland waters, can keep catch rates high in the face of increased fishing and in spite of an ecosystem that is not capable of producing that level of catch through natural processes. Overexploitation can also affect inland fishery resources, but the result is generally a change in species composition and not necessarily a reduced overall catch. Catches are often higher where smaller and shorter-lived species become the main component of the catch; however, the smaller fish may be much less valuable.

Another issue complicating the assessment of inland fishery resources is the definition of a "stock". The major marine fish stocks are well defined biologically and geographically, and comprise management units. Very few inland fisheries have stocks that are defined as precisely or are defined at the level of species. There are notable exceptions, e.g. Lake Victoria Nile perch and Tonle Sap dai fisheries, but many inland fishery stocks are defined by watershed or river and comprise numerous species.

References

Arlinghaus, R. 2006. Overcoming human obstacles to conservation of recreational fishery resources, with emphasis on central Europe. *Environmental Conservation*, 33: 46–59.

Bennett, E., Valette, H.R., Mäiga, K.Y. and Medard, M., eds. 2004. *Room to manoeuvre: gender and coping strategies in the fisheries sector*. Portsmouth, UK, IDDRA. 154 pp

FAO. 2010. *Aquaculture development. 4. Ecosystem approach to aquaculture.* FAO Technical Guidelines for Responsible Fisheries. No. 5, Suppl. 4. Rome. 53 pp.

Gascoigne, J. and Willsteed, E. 2009. *Moving towards low impact fisheries in Europe: policy hurdles & actions.* Brussels, Seas At Risk. 103 pp.

2

Marine Fishery Resources

The marine fisheries sector plays a crucial role for nations across the world. The fisheries resources significantly contribute to food security, income generation, and economic welfare. This resource is an important source for;

— animal protein, many coastal communities consume fish as their main source of protein;

— industry and trade that generates direct and indirect employment; and

— as a source of income to government budgets through fishing agreements, license fees, and from the activities of distant water fishing fleets which are serviced at regional ports.

The more than 6 billion people on Earth consume an average of 15.4 kilograms (almost 34 pounds) of seafood (marine and fresh-water plants and animals) each year. Seafood comes from a combination of capture fisheries (commercial, recreational, or subsistence) and aquaculture. Although the world's population continues to increase, the amount of seafood harvested worldwide has been nearly level since the early 1990s. Many fisheries have declined, and some have collapsed. Fishermen have found a few previously unexploited populations in recent years, but it is clear that the current production level from capture fisheries will not increase in the future. Increases in aquaculture production continue, but the big question is whether production will be able to keep pace with demand.

Traditionally, the status of capture fisheries has been described by (i) providing a summary of the time series trends in production for a fishery or region and (ii) presenting an assessment based on the number of fisheries or stocks categorised as being under-exploited, fully exploited or over-exploited. However, in many seas of the Asia Pacific Region, such descriptions are not sufficient and may in fact be misleading. Reporting total production often masks what is actually happening in the fishery, and both of the above indicators also rely on the existence of accurate and timely fishery statistics from all the sub-sectors of the fishery, including small-scale fisheries. Calculating the percentages in terms of exploitation also relies on having reliable stock assessments, at least for the more abundant fish stocks in a fishery.

The catches in the earlier stages of this development were dominated by large longer-lived predators. As these became fished down, fishers expanded their efforts moving further away from their home base and starting to take smaller, less predatory fish. During this period, total production from the fishery can be expected to be maintained, masking the serial depletion that is occurring. Most alarming is that throughout this period, the total amount of fish in the ocean is continually declining and the catch rates of the major groups are also declining as they, in turn, become overexploited.

The WorldFish Centre (WFC) has recently prepared a number of papers on the status of fisheries from eight APFIC States (Bangladesh, India, Indonesia, Malaysia, the Philippines, Sri Lanka, Thailand and Viet Nam). Overall, analyses of the trawl surveys showed substantive degradation and overfishing of coastal stocks. For trawl surveys where there were more than 25 years between surveys, the amount of fish (measured as tonnes/km^2) had declined to between 6 and 33 percent of the original value. In all cases, the amount of fish had declined, some as much as 40 percent in 5 years. The most dramatic declines were in the Gulf of Thailand and the East coast of Malaysia.

The massive decline in the amount of fisheries resources available for today's fishers has also been associated with changes in the composition of the catch. Surveys in the Gulf of Thailand and in the Lingayen Gulf in the Philippines have shown that the abundance of larger, more valuable species (e.g. groupers, snappers, sharks and rays) higher up in the food chain have declined, while smaller species lower in the food chain (e.g. triggerfish, cardinal fish and squids/octopus) have increased to relative abundance. In the Gulf of Thailand trawl survey catch in the 1960s, the top 10 species (or species groups) were made up of rays, bream, goatfish with some squid, lizard fish and snapper. By contrast in the mid-1990s, less desirable pony fish, squid, barracuda, and lizard fish had become much more prevalent.

Large Marine Ecosystems

The trends in catch composition for the major Large Marine Ecosystems of the APFIC region were examined. LMEs are relatively large regions (200 000 km^2 or more) characterised by distinct bathymetry, hydrography, productivity, and trophically dependent populations. Their seaward limit usually extends beyond the continental shelf. Being defined by natural parameters, they most often straddle political boundaries. They have been identified for the purpose of comprehensive monitoring and could also be used as a basis for ecosystem-based management of shared natural resources in the future.

The Asia-Pacific region is characterised by considerable diversity in LMEs in terms of their fisheries and driving forces. The most productive (total recorded catch since 1950) are the East China Sea, the Kuroshio Current and the South China Sea. Following closely behind, are the Yellow Sea, the Sea of Japan, the NW Pacific high seas area, and the Bay of Bengal.

The Indonesian Sea, the Gulf of Thailand and the Sulu-Celebes Seas are clustered in another group, followed by the much smaller catches in New Zealand and Australia. An interesting feature of many of the LMEs is the dominance of a small number of small pelagic and benthopelagic species in the landings (in terms of weight). The major species are the South American pilchard (Sardinops sagax), chub mackerel (Scomber japonicus) and the largehead hairtail (Triciurus lepturus) in the Kuroshio Current, Sea of Japan, Yellow Sea, East China Sea, and the South China Sea. In the Bay of Bengal, Arabian Gulf, Sulu-Celebes Seas, Indonesian Sea, and NW Australia, anchovies (Stolephorus spp. and Engraulidae) form a large part of the catch with other small pelagic species such as the sardine (Sardinella spp.) common in some areas. Pacific saury (Colobis saira) has formed a large part of the catch in the high seas of the NW Pacific.

Exceptions to this pattern are the heavily trawled areas of the Gulf of Thailand and Northern Australia where more demersals, threadfin breams and shrimps, respectively, are taken, although even in these areas catches of anchovies are almost as high as the demersal catches.

There are obviously numerous ways that the LMEs can be classified:

— Offshore deepwater systems dominated by pelagic fishes (e.g. Kuroshio Current and NW Pacific);

— Heavily fished coastal systems that display sequential depletion of species groups characteristic of "fishing down the food chain" (e.g. Yellow Sea, Gulf of Thailand and NW Australia);

— Coastal systems that are still showing increasing reported catches (e.g. Bay of Bengal and South China Sea); and

— Fisheries managed under tight access right control (e.g. South East Australia and New Zealand Shelf).

Offshore Deepwater Pelagic Systems

The Kuroshio Current LME is dominated by the warm Kuroshio Current that flows in a northeasterly direction along Japan's east coast. This LME has a huge latitudinal expanse, providing it with a rich variety of marine habitats. The region has a generally mild, temperate climate. The underwater topography of the LME includes the Japan Trench, the Shatsky Rise, the Ryukyu Trench and the Okinawa Trough.

The Kuroshio Current LME is considered a moderately high (150-300 gC/m^2/yr) productivity ecosystem with coastal areas that are highly productive. Japan and China are the major fishing nations in the region. The fish catch includes Japanese sardine, Pacific saury, anchovy, jack mackerel, horse mackerel, frigate mackerel, yellowtail, filefish, herring, sea robin, and parrot bass. The sardine population, which is characteristic of small pelagic fish populations, has shown marked fluctuations, attaining an all time maximum in the 1930s, then showing a decrease from 1964 to 1971. It has since increased. With the fluctuations there have been accompanying geographic shifts of spawning and nursery grounds. As is the case in the Sea of Japan, the system has always been dominated by small to mid-sized pelagics, with a more recent trend to large pelagics.

Western and Central Pacific

The Western Pacific Warm Pool is one of the 56 biogeochemical province defined by Longhurst. The Pool is a zone of low productivity which can extend over a range of 80° of longitude (nearly 8 000 km) and has the warmest surface waters in the world. The Warm Pool can undergo spectacular east-west displacements of up to 400 km as part of the El Niño/ La Nina cycle.

The current status of fisheries in the South Pacific was recently reviewed by APFIC. The fisheries catch was divided into oceanic resources that include tunas, billfish that live in the open-water pelagic habitat and coastal fisheries. The offshore resources form the basis of the regions industrialised fisheries and have been fished by an international fleet from 26 different nations over the past 25 years - 15 Pacific Island States and 11 distant-water fishing nations with the bulk of the catch being taken by Japan, USA, Korea RO, and China PR. About 1.6 million tonnes of tuna, as well as an unknown amount of by-catch have been taken from the Western and Central Pacific each year during the 1990s. The main species are skipjack (the majority of which is taken by purse seiners), yellowfin and bigeye tuna (an increasing catch associated with drifting fish attracting devices), and albacore (caught by long-line and trolling with a significant part taken by Pacific Island States).

The coastal resources include a wide range of fin-fish and invertebrates and form the basis for the region's small-scale fisheries. Although dwarfed in both volume and value by the oceanic tuna fisheries, the regions coastal fisheries provide most of the non-imported fish supplies to the region and have a crucial role in food security. The present catch figures are roughly estimated based on agriculture censuses, household surveys or nutrition studies. The best estimate available is about 144 000 tonnes, with about 70 percent of this coming from subsistence fisheries, which despite their importance do not attract much government attention, although anecdotal reports of their depletion in many islands are common.

Major species include finfish, beche-de-mer (sea cucumbers), octopus, lobsters, giant clams, crabs and seaweed. Several high-value products are exported from the region. In a recent World Bank study, coastal fisheries management was examined at 31 locations in the Pacific Islands. The study concluded that there was an urgent need to reduce overall fishing effort. Although many of the communities had adopted restrictions to fishing by outsiders, few were effectively regulating their own harvest.

Ecosystems with "Fishing Down the Food Chain" Effects

Yellow Sea

The Yellow Sea LME is a semi-enclosed body of water bounded by the Chinese mainland to the west, the Korean Peninsula to the east, and a line running from the north bank of the mouth of the Yangtse River (Chang Jiang) to the south side of Cheju Island. It covers an area of about 400 000 km^2 and measures about 1 000 km by 700 km. It is shallow with a mean depth of 44 meters, and it slopes gently from the Chinese continent. The Yellow Sea LME is classified as highly productive (>300 $gC/m^2/yr$) ecosystem. It has marked seasonal variations and supports substantial populations of fish, invertebrates, marine mammals, and seabirds. It has both cold temperate species (eel-pout, cod, flatfish, Pacific herring) and warm water species (skates, gurnard, jewfish, small yellow croaker, spotted sardine, fleshy shrimp, southern rough shrimp).

With its 276 fish species, the Yellow Sea LME is an important global resource for coastal and offshore fisheries. However, it is one of the most intensively exploited areas in the world, and has exhibited a pronounced change in ecosystem structure and "fishing down the food chain" effect. Due to overexploitation and natural fluctuations in recruitment, some of the larger-sized and commercially important species were replaced by smaller, less valuable, forage fish. When bottom trawlers were introduced in the early twentieth century, many stocks were intensively exploited by Chinese, Korean, and Japanese fishers and all the major stocks were heavily fished in the 1960s, which had a significant effect on the ecosystem.

Pacific herring and chub mackerel became dominant in the 1970s. Smaller-bodied and economically less profitable anchovy and scaled sardine increased in the 1980s and took a prominent position in the ecosystem. Cold-water species such as the Pacific cod (Gadus macrocephalus) are almost extinct. It appears fishing has greatly affected both the structure and functioning of the Yellow Sea ecosystem.

Gulf of Thailand

The Gulf of Thailand LME is a semi-enclosed sea immediately to the northwest of the South China Sea LME from which it is separated by two sills. Monsoon seasons and the intrusion of sea water from the South China Sea are the two natural phenomena that seem to drive the LME and are the major causes of oceanographic change in the absence of any massive regime shift. The Gulf of Thailand is relatively shallow, with depths varying between 45 and 80 meters. The LME is considered as being a highly productive (>300 $gC/m^2/yr$) ecosystem. Its high primary production levels are partly the result of increased nutrient loading from rivers and shrimp farms. Primary production is concentrated in coastal areas between Malaysia and Cambodia, and near Viet Nam.

The commercial species found on the shallow Gulf of Thailand consist of crabs, lobsters, rays, sharks and small pelagics (mainly Indian mackerels (Rastrelliger spp.), anchovies, and stolephorus spp.), originally caught by artisanal fishers and supplying local markets. Most important among these are the anchovies used for making fish sauce. In the 1960s, demersal trawl gear was introduced from Germany (as trialed in the Philippines), which led to the development of a Thai demersal trawl fishery, operating in the shallow grounds bordering Thailand's coasts.

The ecological impact of the increase in trawling effort has been well documented and was used to describe the original "fishing down the food chain" phenomenon. The catch composition changed both within species (toward smaller individuals), and between species (toward a mix consisting predominantly of small, short-lived species). The species groups most adversely affected by trawl fisheries were crabs, lobsters, rays, sharks and other large fishes.

Studies of the impact of the various fishing gears operating in the Gulf of Thailand demonstrate that these fisheries have fundamentally altered, and continue to alter, the functioning of that ecosystem. However, even in this extreme case, modelling has shown that the impact still appears to be reversible although a drastic reduction of fishing effort, especially by bottom trawlers and push-netters would be needed to replenish these stocks and halt further ecological degradation of this LME.

Northwest Australian Shelf

The Northwest Australia LME extends from Northwest Cape in the State of Western Australia to the vicinity of the Timor Sea. The LME has a wide continental shelf and it includes topographical features such as the Exmouth Plateau, the Rowley Shelf and the Sahul Shelf. The tropical waters are warm, and the coast includes reefs and extensive mangrove forests. Tropical cyclones are common seasonal events in this LME. It is considered to be a low productivity (<150 $gC/m^2/yr$) ecosystem. The warm tropical waters are the home of corals, fish, starfish, sponges, turtles and shells.

In the Northwest Australia LME, fish stocks are quite small and the level of endemicity is low, with most species distributed widely in the Indo-West Pacific region. Reef fisheries occur in the Rowley Shoals, a chain of coral atolls at the edge of the LME's wide continental shelf. Demersal species that are fished here include snapper, beam, emperors and lizard fishN. These have historically been fished by foreign fleets that caused widespread habitat destruction with an associated decline in the demersal fish catch and a switch to smaller, less valuable fish. A large part of the area is now closed to pair trawlers and access to foreign fleets has been withdrawn, and there is some evidence that the habitat is recovering. A small domestic trap fishery for demersal fish exists in areas subjected to little trawling. It is thought that there will be an expansion of trap fishing in the two closed areas after the species composition changes induced by trawling are reversed.

Systems with Reported Increasing Catches

Bay of Bengal

The Bay of Bengal LME is located in the tropical monsoon belt and is bounded by Bangladesh, India, Indonesia, Malaysia, Maldives, Myanmar, Sri Lanka and Thailand. The Bay's southern part merges into the Indian Ocean. The LME is strongly affected by monsoons, storm surges, and cyclones. Major rivers (Ganga-Brahmaputra-Meghna, Mahanadi, Godavari, Krishna and Salween) introduce large quantities of silt into the Bay of Bengal during the monsoon season from July to September. The Bay of Bengal LME is considered to be moderately productive (150-300 $gC/m^2/yr$).

Commercial species include anchovies, croakers, shrimp and tuna (yellowfin, big eye and skipjack). Shrimp is a major export earner. FAO data show a steady rise in total landings in all major groups since the 1950s. Heavy fishing is a comparatively recent phenomenon, so that stocks have not been subjected to fishing pressure over a lengthy period of time but there is now increased competition and conflicts between small-scale and large-scale fishers.

There is an alarming increase in cyanide fishing in this LME's coral reefs for the lucrative live food fish markets in Hong Kong and Singapore. Mangroves and estuaries - critical fish spawning and nursery areas - are also under stress or threatened by pollution, sedimentation, dams for flood control (as in Bangladesh), and intensive coastal aquaculture. States bordering the Bay of Bengal rate overexploitation of marine resources as the number one problem in the area.

South China Sea

The South China Sea LME is bounded by the coasts of Viet Nam, China

PR, Taiwan POC, the Philippines, Malaysia, Thailand, Indonesia and Cambodia. It is separated from the Gulf of Thailand to the West, by a shallow sill. The South China Sea contains many biological subsystems and a variety of habitats. These include mangrove forests, seagrass beds, coral reefs and soft-bottom communities. The 50-meter depth contour largely follows the coast, with the widest shelves occurring along the eastern edge of the LME. Much of the South China Sea is below 200 meters with oceanic waters, ranging in depth from 200 to 4 000 meters, covering nearly half of the Sea.

It is considered a moderately high productivity (150-300 gC/m^2/yr) ecosystem. High productivity levels are found in gulfs, along the coast, and in reef and seagrass areas, common in the Philippines portion of the LME. The coastal and estuarine areas off Viet Nam, China PR and Cambodia are very productive and, in the past, a substantial fraction of the catch was taken by artisanal, non-mechanised boats. The Viet Nam/China PR area was lightly exploited from the mid-1970s to the mid-1980s, but by now much of this potential has probably been realised.

The total fish harvest was approximately 5.0 million tonnes a year in 2001, with an increasing trend until very recently. Although catches were apparently increasing for all trophic level groups, ecosystem modelling has shown that this area has also been subjected to "fishing down the food chain". Five of the States fishing this LME are among the top eight shrimp producers of the world. Fishermen sometimes use small-meshed nets and practice destructive fishing methods, such as cyanide and dynamite fishing.

In deep oceanic waters (200 to 4,000 m), fisheries are limited mainly to large pelagic fishes - tuna with some billfish, swordfish, shark, porpoise, mackerel, flying fish and anglerfish. The deeper coralline areas and those situated in the central portion of the LME are only lightly exploited, leaving room for a possible increase in catch from this area, although the resources are probably rather limited.

Fisheries Under Tight Management Control

Southeast Australian Shelf

The Southeast Australia LME extends from Cape Howe, at the southern end of the state of New South Wales, to the estuary of the Murray-Darling river system in the state of South Australia. It borders the Southern Ocean and the western boundary currents flowing into the West Wind Drift, which circulates around the continent of Antarctica. It contains the island of

Tasmania and Bass Strait, which separates that island from the state of Victoria on the mainland. The LME has a diversity of habitats such as seagrass beds, mud flats, intertidal and sub-tidal rocky reefs, mangrove forests and pelagic systems. It has been classified as a highly productive (>300 gC/m^2/yr) ecosystem, despite the low nutrient input into the area.

Some of the species harvested are scallops (in Bass Strait), rock lobster (Tasmania), and abalone (Tasmania and Victoria) and finfish in the Southeast trawl fishery. The long-standing trawl fishery has seen serial depletion of several of its important fish stocks, including eastern gemfish and more recently the deepwater orange roughy that is found associated with sea mounts. The fishery has been managed under individual transferable catch quotas (ITQ) since the early 1990s, but to date, these do not appear to have been very effective in reducing overfishing in this multi-species fishery.

New Zealand Shelf

The New Zealand Shelf LME surrounds the islands of New Zealand and stretches across 30 degrees of latitude from the sub-tropics in the north down to the sub-Antarctic region. The shelf surrounding New Zealand varies in width from 150 km in the northeast and southwest, to 3 000 km on the northwest and southeast plateaus. The northern half of the LME is influenced by the warm South Equatorial Current, while the southern half is influenced by the cooler West Wind Drift. The marine environment is diverse and includes estuaries, mudflats, mangroves, seagrass and kelp beds, reefs, seamount communities and deep sea trenches. The New Zealand Shelf LME is considered to be a moderately high (150-300 gC/m^2/yr) productivity ecosystem.

Maori cultural ties with fisheries are strong and their fishing rights are recognised in law. Fishing is a popular leisure activity for as many as one in five New Zealanders. Among the important commercial fisheries of the area are those for migratory predators such as tuna, billfish, and sharks. There is also a bottom fishery for the orange roughy, an important blue grenadier fishery and a coastal fishery for a variety of crustaceans and mollusks. The fishing industry is mostly export-oriented. All fisheries are managed under an ITQ system, that the NZ government claims has resulted in sustainable fisheries. Prior to the introduction of the system, overfishing of coastal resources had occurred. As in Australia, the long-lived deepwater species such as orange roughy have continued to decline, offset by the discovery of new stocks and expansion in fishing. Other important stocks, such as the blue grenadier appear to be fished sustainably at present.

Inland Waters

Water Resources

Asia is blessed with the most freshwater in the world, estimated at 13 510 km^3. However, because of the high population density, the per capita availability of freshwater is the lowest and competing usages of freshwater has a major impact on fisheries. The main water resources for fisheries are the rivers and flood plains, natural lakes and man-made impoundments. Asia has the largest number of rivers than any other continent and also has the highest cumulative river channel length. The seasonal floodplains of States such as Myanmar and Bangladesh all have major fisheries. In contrast, Asia has relatively few natural lakes. Most of these are confined to volcanic areas, notably in the Indonesian and Philippine archipelagos, and in northern India. The major exception is the Great Lake or Tonle Sap in Cambodia which occupies 200-300 km^2 in the dry season, expanding to 10 000-12 000 km^2 in the wet.

Reservoirs have been constructed over many centuries in some States in Asia, (e.g. Sri Lanka), but more recently, construction of reservoirs for irrigation, hydropower and flood protection has become common in many parts of the Asia-Pacific region. These come in many forms ranging from large impoundments resulting from damming major rivers to small rain-fed ponds. The reservoir resource in Asia is large and accounts for over 40 percent of the global large reservoir capacity. It has been estimated that developing States in Asia have 66.7 million ha of small to medium reservoirs with 85 thousand ha occurring in China.

Fishery Resources

Most inland fisheries are small-scale activities where the catch per craft (or catch per capita) is relatively small and the catch more often than not disposed of on the same day. The main exceptions are the industrialised fisheries in the lower Mekong Basin and the "fishing lots" in the Tonle Sap of Cambodia and the fishing "inns" of Myanmar. The lack of accurate reporting of these small-scale fisheries makes it difficult to describe their status but it is generally felt that they are under considerable pressure from loss and degradation of habitat and overfishing. Freshwater fishes are reported to be the most threatened group of vertebrates harvested by man. The World Resources Institute estimated that half of the World's species were lost during the last century and that dams, diversions and canals have fragmented many major rivers, severely impacting fisheries resources.

In general, the reported catches of inland fisheries in most States have continued to increase, although many of the inland water habitats have been altered and degraded. Rivers and floodplains, in particular, have been heavily impacted with the construction of dams, roads, channels and other irrigation systems. The training of water to reduce the impacts of flooding has probably had widespread effects on inland fisheries in the region, although the real impacts are poorly documented. Fishers regularly complain that catches are declining, but it is uncertain to what extent this is an effect of increasing fishery pressure or the loss of fishery resources through habitat degradation and changing water flow regimes.

In some areas, the way the systems function has been altered enormously by greatly reducing the seasonal "flood pulse" that previously resulted in large areas of land becoming flooded during the wet season. This in turn has impacted on the ability of many species of fish to migrate to their spawning grounds and has also altered the flow pattern with accompanied habitat changes (e.g. running water to lakes). Some mitigation attempts have been made in terms of fish ladders to allow fish to migrate around dams but their success compared with original conditions is typically unknown.

Changing agricultural cropping patterns, irrigation development and the increased use of chemicals and pesticides in intensified agricultural production have all had some impact on wild fishery resources. Changing rural livelihoods have also led people away from their traditional dependence on fishery resources as livestock has supplanted wild resources.

As in the marine environment, changes in species composition are also occurring. Catches of large, long lived species such as the giant catfish and large cyprinid species are becoming rare and there is evidence of "fishing down the food chain" in some inland areas. In the Mekong River, for example, the giant catfish (Pangasianodon gigas) has now become extremely rare and endangered. On the other hand, the status of some inland fishery resources has been enhanced through stocking programmes, introductions of exotic species, habitat engineering and habitat improvement.

Stock enhancement is an integral part of most inland fisheries in the Asia-Pacific region. With developments in artificial propagation techniques of fast-growing and desirable species and increased availability of seed stock, the use of this intervention is increasing, particularly where some sort of access restriction or user rights is in place. The economic viability of stock enhancement in large lakes and rivers has not been demonstrated in any Asian State and, in general, these fisheries depend naturally on recruited stocks. In floodplain depressions and in small water bodies, stock enhancement has proved to be very successful. These cultured-based fisheries are seen as a way forward in China PR and many States in the region.

Of particular importance to many States of the Asia-Pacific region are rice-field fisheries that either depend on natural introduction of wild fish or stocking fish, either simultaneously or alternately with the rice crop (probably more correctly defined as aquaculture). This aquatic production, in addition to the rice crop itself, is important for rural livelihoods in developing States. Its local consumption and marketing are particularly important for food security as it is the most readily available, reliable and cheapest source of animal protein for farming communities as well as the landless.

This importance is generally underestimated and undervalued. It is reported that the availability of this aquatic resource is declining. Anecdotal evidence from China, Viet Nam, Lao PDR, Cambodia, Thailand and elsewhere, suggests that it is considerably more difficult to find such food now than a decade or so ago. This is a result of increasing demand through human population increases and activities such as the use of pesticides, destruction of fish breeding grounds and the use of illegal fishing methods such as poisons and electro-fishing.

Production and Trends in Marine Fisheries

A major attempt to assess the global state of the world's marine fishery resources and to estimate their long-term production in the late 1960s and early 1970s. In those years, while the total world marine capture fish production was approaching 60 million tonnes per year after having increased at a rate of 6 percent per year since 1950.

Total world fish production increased steadily from 19.3 million tonnes in 1950 to more than 100 million tonnes in 1989 and 134 million tonnes in

2002. Marine capture fisheries are the largest contributors to world fish production. In 1950 marine captures were 16.7 million tonnes, representing 86 percent of the total world fish production, and by 1980 marine captures increased to 62 million tonnes, also representing 86 percent of the total.

However, over the last two decades there has been a faster expansion of marine and inland water aquaculture, and the relative contribution of marine capture fisheries to the total world fish production has diminished. Nevertheless, total catches had continued to increase, but at a slower rate than aquaculture. At present, out of the total world fish production of 134.3 million tonnes in 2002, almost 63 percent (84.4 million tonnes) were produced through the exploitation of wild fish resources of the oceans. Marine and inland water aquaculture represent around 30 percent and inland water capture fisheries represent the remaining 7 percent.

After reaching about 80 million tonnes in the late-1980s, global marine catches fluctuated between 77 and 86 million tonnes, with a record high of 86.7 million tonnes in 2000 and a slight decline to 84.4 million tonnes in 2002. Most of the fluctuations in recent years were the result of changes in a few highly productive areas, particularly in the northwest Pacific and the southeast Pacific, while total catches in the majority of the other fishing areas followed more or less the same general trend. Although in some areas catches have started to level off since the 1970s, such as in the northeast Atlantic, western Central Atlantic, eastern Central Atlantic, Mediterranean and Black Sea, northwest Pacific and eastern Central Pacific.

In other areas, such as the northwest Atlantic and the southeast Atlantic, catches reached a maximum in the 1970s or early 1980s and are declining. The general trend described above provides support to the earlier estimates by Gulland, and it suggests that the maximum long-term potential of the world marine capture fisheries has been reached, with some stocks and areas being overfished and some stocks not producing their full expected long-term potential.

Marine Catch Composition

The larger portion of the global marine catches are pelagic species, with small pelagics representing around 26 percent (22.5 million tonnes) of the total catch in 2002, down from 29 percent in the 1950s and 27 percent in 1970s. The larger pelagics accounted for 21 percent (17.7 million tonnes) of the total catches in 2002, an increase in their share from 13 percent in the 1950s. Demersal fishes contributed 15 percent of the total catches in 2002 (with 12.3 million tonnes), compared with almost 26 percent of the world catches in the 1950s and 1970s.

Miscellaneous coastal fishes remained stable at 6 percent and then 7 percent (with 6.1 million tonnes) in 2002, while crustaceans increased from 4 percent in the 1950s and 1970s to 7 percent (5.8 million tonnes) in 2002. Molluscs increased slightly from 6 percent in the 1950s and 1970s to 8 percent (6.8 million tonnes) in 2002. There was a slight increase in the proportion of unidentified fish, in 2002 with 13 percent of the total catches (10.7 million tonnes), up from 11 percent in the 1950s and 1980s.

All of the above major species groups are represented more or less equally in the northwest Pacific (Area 61), the most productive fishing area of the world. Small pelagics (mostly Peruvian anchoveta) clearly dominate catches in the southeast Pacific, the second most productive area. In the northeast Atlantic, the third most productive area, demersal fishes are the most abundant, followed by larger pelagics and small pelagics.

In the western Central Pacific, the fourth most productive area of the world, catches are dominated by larger pelagics, which are also the most abundant group in the western Indian Ocean. Small pelagics are also dominant in the eastern central Atlantic, Mediterranean and Black Sea, western central Atlantic and eastern central Pacific, while demersal fishes are the dominant species group in the northeast Pacific and southwest Pacific.

Catch Fluctuations

Catches in the northwest Pacific have oscillated between 20 and 24 million tonnes since the late 1980s, with the larger fluctuations being caused by catch and presumably abundance changes of Japanese pilchard or sardine (*Sardinops melanostictus)* and Alaska pollock (*Theragra chalcogramma*). In the southeast Pacific, only three species account for more than 80 percent of the current and historical catches. These are the Peruvian anchoveta (*Engraulis ringens*), the Chilean jack mackerel (*Trachurus murphyi*) and the South American pilchard or sardine (*Sardinops sagax*) that have had alternating periods of high and low abundance over the past decades. Large catch fluctuations are common in this area and are mostly a consequence of the periodic climatic events known as El Niño, affecting fishing success as well as longer term stock abundance and productivity.

Important changes are also reported for other regions of the world, although their combined effect on global catches is less noticeable. In the western Central Pacific total catches have been increasing steadily since 1950, reaching almost 11 million tonnes in 2002. In the northwest Atlantic fish catch production declined to a low of 2 million tonnes in 1994, following the collapse of groundfish stocks off eastern Canada. Since then catches have increased slowly to 2.3 million tonnes in 2002.

Tunas and tuna-like species are the most important fishery resources exploited in the high seas. Their production is considerably higher in the Pacific Ocean followed by the Atlantic and Indian Oceans. A recurring pattern in some areas is the medium to long-term change in catch composition following the decline of some fish stocks that traditionally have been dominant in the area. For instance, in the northwest Atlantic catches of molluscs and crustaceans have increased noticeably following the declines of demersal fishes. In the northeast Atlantic the reduction in catches from the continuous decline of Atlantic cod (*Gadus morhua*) since the late 1960s has been balanced out by the increase in catches of formerly low-value species, such as blue whiting (*Micromesistius poutassou*) and sandeels (*Ammodytes spp.*).

The severe decline of the Argentine hake (*Merluccius hubbsi*) in the southwest Atlantic has been accompanied by an increase in the catch of shortfin squid (Illex argentinus). In the northwest Pacific the decline in catches of the Japanese pilchard or sardine (Sardinops melanostictus) and the Alaska pollock (*Theragra chalcogramma*) has been somewhat compensated by the increasing catches of the Japanese anchovy (*Engraulis japonicus*), the largehead hairtail (*Trichiurus lepturus*) and squids (*mostly Todarodes pacificus*).

The causes for these medium to long-term changes in the species composition of marine commercial catches can be multifold, including the adaptation of industry and markets to previously unattractive low valued species, the effect of fishing on the abundance of target species and on the structure of other marine communities, as well as environmental changes or regime shifts effecting the long-term abundance of the various wild fish stocks. Often these effects are confounded and in many cases they are difficult to discern, particularly in areas where research and monitoring of fishery resources and environmental processes are not well developed.

Exploitation of Fishery Resources

Some fishing areas of the world there is a relatively large number of stock or species groups whose state of exploitation is undetermined or not known. The stocks of seven of the top ten species that account for 30 percent of the world total marine capture fisheries production are either fully exploited or overexploited and therefore no sustainable increases in catches can be expected from these species. These include two stocks of Peruvian anchoveta (*Engraulis ringens*) in the southeast Pacific that are overexploited after recovering from a recent decline, Alaska pollock (*Theragra chalcogramma*) that is fully exploited in the North Pacific, Japanese anchovy (*Engraulis japonicus*) that is fully exploited in the northwest Pacific, blue whiting (*Micromesistius poutassou*) that is overexploited in the northeast Atlantic, capelin (*Mallotus villosus*) that is fully exploited in the North Atlantic and Atlantic herring (*Clupea harengus*) with several stocks in the North Atlantic, most of them fully exploited.

Another two of the top ten species can probably support some limited increase in catches in parts of their distribution range where they are reported as still moderately exploited. This is the case of the chub mackerel (*Scomber japonicus*) in the southeast Pacific and eastern Central Pacific and Skipjack tuna (*Katsuwonus pelamis*) in the Pacific and Indian Oceans as well as in parts of the western Atlantic. In other parts of their worldwide distribution range these two species are also reported as fully exploited, and therefore offer no possibilities of a sustained increase in catches. The status of the tenth top species, the largehead hairtail (*Trichiurus lepturus*) is reported as unknown in most of its areas of distribution, particularly in the northwest Pacific and western Indian Ocean. In other areas such as the eastern Indian Ocean and western Central Atlantic it is far less abundant.

The major fishing areas with the highest proportions (69-77 percent) of fully exploited stocks are the western Central Atlantic, eastern Central Atlantic, northwest Atlantic, western Indian Ocean and northwest Pacific. While the areas with the highest proportions (46-60 percent) of overexploited, depleted and recovering stocks are the southeast Atlantic, southeast Pacific, northeast Atlantic and (for tuna and tuna-like species) Oceanic areas of the Atlantic and Indian Oceans. Few areas of the world report a relatively high number (48-70 percent) of species or stock groups as still being underexploited or moderately exploited, as is the case of the eastern Central Pacific, western Central Pacific and southwest Pacific. Some areas have 20-30 percent of stocks still considered moderately or underexploited such as the Mediterranean and Black Sea, southwest Atlantic and eastern Indian Ocean.

The combination of heavy fishing pressure and severe adverse environmental conditions associated with changes in the El Niño Southern Oscillation led in recent years to the sharp decline in the three most abundant species in the southeast Pacific, the Peruvian anchoveta, the South American pilchard (or sardine) and the Chilean jack mackerel. The stocks of Peruvian anchoveta have shown signs of recovery and at present are most likely fully or overexploited with catches in the order of 7 to 11 million tonnes per year after a sharp decline to only 1.7 million tonnes in 1998.

The South American pilchard has declined sharply as part of a decadal regime period and in 2002 yielded only 28 000t after yielding up to 6.5 million tonnes in 1985 during its latest high abundance regime period. The Chilean jack mackerel is assessed as fully to overexploited and yielded 1.7 million tonnes in 2002 after declining continuously from a peak production of 5 million tonnes in 1994. The Chilean jack mackerel and particularly the South American pilchard are at present in a period of natural low abundance and it is expected that they may enter a period of high abundance when favourable environmental conditions return, provided they are not excessively exploited during the current phase. It is also unlikely that environmental conditions would be favourable for all three species above simultaneously.

In the northwest Pacific large changes in the abundance of Japanese pilchard or sardine, Japanese anchovy and Alaska pollock have also occurred in response to heavy fishing and to natural decadal oscillations. After a high abundance period in the 1980s, the Japanese pilchard declined in the mid-1990s and was followed by a strong recovery of the Japanese anchovy population which has supported catches close to 2 million tonnes per year since 1988. This alternation of sardine (or pilchard) and anchovy stocks follows a pattern also observed in other regions of the world that seem to be mainly governed by climatic regimes affecting stock distribution and overall fish abundance.

The stocks of Alaska pollock in the northwest Pacific are considered to be fully to overexploited, while those in the northeast Pacific are considered fully exploited. Catches of Alaska pollock peaked in the late-1980s and have been declining since in both areas, although there is a rècent recovery in the northeast Pacific. In the northeast Atlantic catches of blue whiting have increased steeply and the species is considered overexploited. Most of the stocks of Atlantic cod in the area are also overexploited or depleted, while capelin and herring are exploited to their full potential

There is a long way to go before society can be satisfied with the trends and state of exploitation of world fisheries resources. Almost 76 percent (52 percent that are fully exploited and 24 percent that are or have been overexploited) of fish stocks for which assessment information is available, need to be monitored and managed on a continuous basis and/or need to be rebuilt to ensure sustainability objectives.

Therefore greater efforts are to be made at all levels and stages of fisheries research and management, and these should range from improving the identification of species being caught and landed, to improving the information base for the proper assessment of fish stocks and management of their fisheries. These are some of the most important steps to effectively promote the application of the Code of Conduct for Responsible Fisheries and assist member countries in complying with the recommendations of the 2002 Johannesburg World Summit on Sustainable Development regarding the need to rebuild overexploited and depleted stocks with specific targets for the year 2015. It is hoped that this review will contribute towards these aims.

References

Adams, T., "Coastal fisheries and marine development issues for small islands", In M.J. M. J. Williams, ed., *A roadmap for the future for fisheries and conservation, ICLARM Conference Proceedings* No. 56: 40 - 50, 1998.

Bell, J.D., "Transfer of technology on marine ranching to small island states", *In (unedited) Marine ranching: global perspectives with emphasis on the Japanese experience, FAO Fisheries Circular no.*, 943: 53 - 65, 1999b.

Dalzell, P., Adams, T.J.H. & Polunin, N.V.C., "Coastal fisheries in the Pacific Islands", *Oceanogr. Mar. Biol.: an Annual Review*, 34: 395 - 531, 1996.

MacCall, A.D., "Against marine fish hatcheries: ironies of fishery politics in the technological era", *Cal. COFI Reports*, 30: 46 - 48, 1989.

McCarty, C.E., McEachron, L.W. & Rutledge,W.P., "Beneficial uses of marine fish hatcheries: the Texas experience", *The International Symposium on Sea Ranching of Cod and Other Marine Species*, Arendal, Norway, 15 - 18 June 1993, Programme and Abstracts, 1993.

3

Methods of Fish Capture

Capture fisheries resources are highly diverse. By far the most numerous fish species, and those most important to aquaculture and fisheries, are teleosts, or bony fish, which in the sea extend from small grazing species such as anchovy to large active predatory fish such as tuna. A similarly wide range is also found in freshwater, with the most important species from a production point of view belonging to the carp family. These account for over half the total of inland waters fisheries production.

Marine capture fisheries resources are usually considered close to full exploitation worldwide with about half of them fully exploited, one quarter over exploited, depleted or recovering from depletion and one quarter only with some capacity to produce more than they presently do. The overall situation of inland capture fisheries resources is not as well known but is likely to be as serious or worse, considering the much larger environmental impact they are subject to.

Capture fisheries resources are usually exploited and managed on a stock-by-stock basis. Stocks present a wide range of characteristics that affect the fisheries exploiting them: their mono- or multi-species composition, size, value and distribution. High seas resources such as tuna or marine mammals require international collaboration for their management.

Types of Capture Fisheries

Capture fisheries are extremely diversified, comprising a large number of types of fisheries that are categorised by different levels of classification.

On a broad level, capture fisheries can be classified as industrial, small-scale/ artisanal and recreational.

A more specific level includes reference to the fishing area, gear and the main target species, such as the North Sea herring purse seine fishery, Gulf of Mexico shrimp trawl fishery, southern ocean Patagonian toothfish longline fishery. While capture fisheries encompass thousands of fisheries on a global scale, they are often categorised at the level of which a fishery is managed nationally and/or regionally.

Industrial Fisheries

Capital-intensive fisheries using relatively large vessels with a high degree of mechanisation and that normally have advanced fish finding and navigational equipment.Harvesting fish for personal use, leisure, and challenge (e.g. as opposed to profit or research). Recreational fishing does not include sale, barter or trade of all or part of the catch. Fisheries undertaken for profit and with the objective to sell the harvest on the market, through auction halls, direct contracts, or other forms of trade.

A fishery where the fish caught are shared and consumed directly by the families and kin of the fishers rather than being bought by intermediaries and sold at the next larger market. Pure subsistence fisheries are rare as part of the products are often sold or exchanged for other goods or services.

Fisheries established long ago, usually by specific communities that have developed customary patterns of rules and operations. Traditional fisheries reflect cultural traits and attitudes and may be strongly influenced by religious practices or social customs. Knowledge is transmitted between generations by word of mouth. They are usually small-scale and/or artisanal.

Some industrial fisheries are quite easily identified. For instance, the world's largest fishery—the Peruvian purse seine fishery for anchoveta—is one species caught mainly by one country in a specific area. However, the anchoveta do coexist with other species such as mackerel and the resource can move south into Chilean waters. The resource also has great fluctuations caused by "El Niño", so large that they are even seen in the total world's catch figures. The schooling characteristics of pelagic fish such as anchoveta do mean that that they are generally mono-species, which more readily identified as a "fishery".

At the other end of the scale, the North Sea demersal fishery consists of about 20 species caught by some 12 countries using a variety of fishing gears. It is possible to disaggregate this major fishery into sub-fisheries, by country, species or area, but none of these would be a cohesive fishery unit because other sub-fisheries would impact it. This amalgam of sub-fisheries makes them very difficult to manage. The range of different fishing gears and methods generates conflicts among fishers, often with the most efficient gear blamed for overfishing and/or damage to the environment.

This range of fishing methods is, in many cases, a measure of the fishery's maturity given that, as each more efficient method is introduced, a remnant of the previous prevailing fishing method still co-exists. The wide range of target species gives rise to one of the most intractable problems of fisheries management: that of the optimum fishing effort on a multi-species fishery. Each species has a different optimum fishing effort and the management decision on the fleet's optimum will be greater than the optimum for any one species. This leads to overfishing on some species and a less than optimum catch for others. Attempts to correct this maladjustment by imposing catch quotas cause low value fish to be returned in order to keep more of the higher value species—thus optimising the boat's revenue when the overall catch quota is close to being reached.

Given these two different situations, it is easy to classify up to about 70 percent of the world's catch into around 200 fisheries. However, the remaining 30 percent is simply classified as the North Sea fishery (an Area classification) broken down into species and/or countries. As a general rule, industrial fisheries are confined to the continental shelf where fish are more abundant, although there are exceptions.

The fisheries for large pelagics such as tuna and swordfish are ocean-ranging using purse seines and longlines. However, the global total catch of these vessels would be only around five percent of the those vessels fishing on the continental shelf. Another group of fishing vessels presently attracting much attention target Patagonian Toothfish and similar species, but again their numbers are minimal compared to the continental shelf or near coastal fisheries.

The most spectacular industrial fisheries were the fleets of factory vessels and factory trawlers built in the Soviet bloc between 1960—1980. These very large vessels fished all around the world and were supported by a fleet supplying fuel and provisions and transporting the catch. The largest vessels were about 8 000 GRT and more than 120 metres in length. The extension of the Exclusive Economic Zones (EEZs) in the early 1980s restricted the fishing grounds that were freely available to these vessels but in many cases fishing agreements between the flag State and the coastal State allowed the vessels to continue fishing.

As coastal countries developed the ability to harvest their own resources, such agreements were more difficult to negotiate and, around 1990, the building programme for these very large vessels came to an end. Although there are still some very large vessels currently being built, these are exceptions and the numbers are in single figures compared to the hundreds built in the 1970s and 1980s.

Small-scale Fisheries

Small-scale fisheries also referred to as artisanal fisheries, are difficult to define unambiguously, as the term tends to apply to different circumstances in different countries. In general, they are traditional fisheries involving fishing households (as opposed to commercial companies), using relatively small amounts of capital and energy, relatively small fishing vessels (if any), making short fishing trips close to shore, mainly for local consumption.

In practice, the definition varies between countries, e.g. from gleaning or a one-man canoe in poor developing areas, to more than 20-m. trawlers, seiners or long-liners in developed ones. Artisanal fisheries can be subsistence or commercial, providing for local consumption or export. As well, while most artisanal fisheries produce fish that is shared and consumed directly by the fisher's kin rather than being bought by middle-traders and sold at the next larger market, some are truly export-oriented (e.g. cephalopod pot fisheries in Mauritania). However, it can probably be said, that pure subsistence fisheries are rare, as part of the products are very often sold or exchanged for other goods or services.

Artisanal Fisheries

Artisanal fisheries can be very specialised but, in general, target a very wide range of species, using a broad variety of gears, generating diverse fishing strategies and flexibly adapting to seasonal or inter-annual natural variability. Artisanal fisheries are important for many reasons. Globally, they produce about 50% of the world capture fisheries' harvest used for human consumption and as such contribute significantly to food security, particularly in rural areas. They also provide significant employment.

Although available statistics are extremely poor, FAO estimates indicate that there were close to 29 million fishers—artisanal and industrial—in 1990 and that this number has been holding steady since. Artisanal fishers represent the lions' share—a guestimate would put the figure at around 20 million. In addition, downstream industries and support services generate possibly another 80 million jobs, ensuring some livelihood for 200 million people (assuming a ten-to-one ratio).

Locally, artisanal fisheries often provide an economic activity and livelihood of last resort for the poorest strata of the rural—and even sometimes urban—populations. In cases of exceptional conditions, such as severe droughts in Africa, they may be the only occupation possible for displaced peoples. Close to urban centres, they often provide a livelihood for the jobless. This is made possible both by the traditional cultures of mutual assistance as well as by the free and open access nature of many of the small-scale fisheries.

Access to small-scale fisheries is most often neither limited nor really controlled by central fisheries management authorities. However, in many areas, traditional regulations and relationships exist by initiating access control through local ethnic groups or communities such as fishing fees by foreigners and the right to establish a fishing camp.

Artisanal fisheries are difficult to administer because they are largely scattered along the edges of aquatic systems, rivers, lakes and marine shores, including difficultly- accessible areas. This characteristic explains the severe constraints faced by artisanal fisheries in terms of management, access to modern technology, capital, health care, markets, electricity, education, manpower, etc. These constraints are compounded by the lack of mobility (out of the sector and the area) and the fact that many small-scale fishers are also part-time farmers.

Contrary to the impression often obtained from a superficial visit, artisanal fisheries are often extremely dynamic. In some countries (e.g. China, Guinea) they have recently expanded rapidly. In others (Senegal, Côte d'Ivoire) they are stagnating after a very vigorous expansion the past two decades. In others still (Japan, Malaysia), the number of artisanal fishers is declining, following general trends in demography, economic development and availability of better paid jobs, or as a consequence of urban drift.

Artisanal fisheries technology is also evolving rapidly with the adoption of modern materials such as fibreglass boats, multimonofilament nets, outboard and in-board engines, echo sounders, satellite global positioning systems (GPS).The role of women in artisanal fisheries is particularly important, especially in the areas of fish processing, distribution, wholesale and retail marketing. These activities contribute substantially to the maintenance and development of the household. Intermediaries play a major role in the development of artisanal fisheries as fishmongers traditionally control the market outlets and represent the only source of credit available for very cash-strung fishers. Interest rates are inclined to be high and, despite the reality of the grave financial risks taken by these informal bankers, tend to be considered excessive.

Migration is a common characteristic of many artisanal fisheries. Fishers often move seasonally to follow fish across their migratory routes, inside a country (e.g. between north and south Senegal) or between countries (e.g. Senegal and Mauritania; Ghana or Mali and Côte d'Ivoire). They may also move more permanently, creating quasi-permanent settlements abroad (e.g. Ghanaian fishers found all along the Atlantic coast of Africa).

Artisanal fisheries often fare well when compared to their more commercial or industrial counterparts. They use much less fuel and generate more employment per tonne of fish produced. Their economic performance, often underestimated and sometimes ignored by national governments, can be very high, particularly considering that this sector receives very few subsidies, if any.

While most contribute to local markets, some are actively export-oriented towards regional or international markets (e.g. Senegalese processed fish to the Gulf of Guinea; South African abalone to Asia; Northwest African high-quality fish and octopus to Japan or Europe).

Their environmental impacts on fish stocks and fish habitats are often less than that of commercial fisheries largely because of the widespread use of selective and stationary fishing gear, as well as the overall lower fishing power exerted than in industrial fisheries. However, this statement should be qualified, on several counts. Firstly, there is the fairly widespread use of destructive practices such as poison (cyanide) and explosives (dynamite and other "home-made" explosives) in some small-scale fisheries, especially in tropical reef areas. Secondly, some artisanal fishing gear are very unselective, such as small-meshed beach-seines and some types of trammel, lift and cast nets . Thirdly, the sheer density and intensity of fishing off highly populated coastal areas can contribute to serious levels of overexploitation of vulnerable fish stocks such as long-lived demersal resources.

Practically everywhere, artisanal fisheries tend to be confronting severe competition from commercial fisheries for fishery resources and fishing areas, markets, financial support from governments. Conflicts are spreading with significant consequences for household economies (such as gear destruction, market losses) and, sometimes, peoples' life. In the past, these conflicts have been very violent in some areas (e.g. Southeast Asia). Overall, the political power of the small-scale sector depends on its size, degree of organisation (within a powerful association, for example) and ethnic relationships with political leaders.

Fish Capture Technology

Fish capture technology encompasses the process of catching any aquatic animal, using any kind of fishing methods, normally operated from a vessel. Use of fishing methods varies, depending on the types of fisheries, and can range from a simple and small hook attached to a line to large and sophisticated midwater trawls or purse seines operated by large fishing vessels. The targets of capture fisheries can include aquatic organisms from small invertebrates to large tunas and whales, which might be found anywhere from the ocean surface to 2,000 meters deep.

The large diversity of targets in capture fisheries and their wide distribution requires a variety of fishing gear and methods for efficient harvest. These technologies have developed around the world according to local traditions and not least technological advances in various diciplines.

In recent decades major advances in fiber technology, along with the introduction of other modern materials, have made possible, for example, changes in the design and size of fishing nets. The mechanisation of gear handling has vastly expanded the scale on which fishing operations can take place. Improved vessel and gear designs, using computer-aided design methods, have increased the general economics of fishing operations. The development of electronic instruments and of fish detection equipment has led to the more rapid location of fish and the lowering of the unit costs of harvesting. Developments in refrigeration, ice-making and fish processing equipment have contributed to the design of vessels capable of remaining at sea for extended periods.

While these technologies are available, those actually introduced in many small-scale fisheries may amount to no more than motorising a dugout canoe, use of modern and lighter gear or introducing the use of iceboxes to ensure the quality of the product landed. The impact of such changes, however, has considerably increased landings and the earnings of fishers, and underlines the need for effective management to prevent excessive fishing effort. The emphasis of much recent technical innovation has been focused on greater selectivity of fishing gear and on gear with less impact on the environment.

Fishing Gears and Methods

Methods to catch fish and other aquatic resources, with or without a gear, have always been practiced. Although the fundamental principles, i.e. filtering the water, luring and outwitting the prey and hunting, are the basis for most of the fishing gears and methods used even today, gears and methods have changed significantly over time and their capture efficiency is obviously hardly comparable to that of prehistoric times.

A fishing gear is the tool with which aquatic resources are captured, whereas the fishing method is how the gear is used. Gear also includes harvesting organisms when no particular gear (tool) is involved. As well, the same fishing gear can be used in different ways. A common way to classify fishing gears and methods is based on the principles of how the fish or other prey are captured and, to a lesser extent, on the gear construction. FAO defines and classifies the main categories of fishing gear as follows:

- surrounding nets (including purse seines);
- seine nets (including beach seines and Boat, Scottish/Danish seines);
- trawl nets (including Bottom: Beam, Otter and Pair trawls, and Midwater trawls: Otter and Pair trawls);
- dredges
- lift nets;
- falling gears (including cast nets);
- gillnets and entangling nets (including set and drifting gillnets; trammel nets);
- traps (including pots, stow or bag nets, fixed traps);
- hooks and lines (including handlines, pole and lines, set or drifting longlines, trolling lines);

— grappling and wounding gears (including harpoons, spears, arrows, etc.);

— stupefying devices.

Fishing methods have continuously evolved throughout recorded history. Fishers are inventive and not afraid of trying new ideas. The opportunities for innovation have been especially good in recent decades with advances in fibre technology, mechanisation of gear handling, improved performances of vessels and motorisation, computer processing for gear design, navigation aids, fish detection just to mention a few.

Whereas technological development of fishing gear and methods in the past was aimed at increasing production, the present situation with many overfished stock, limited possibilities to expand fishing on underexploited resources and concerns about the environmental impacts of fishing operations, gear development is now very much focused on selective fishing and gears with less impact on the environment.

Categories of Fishing Gears

A fishing gear is the tool with which aquatic resources are captured, whereas the fishing method is how the gear is used. Gear also includes harvesting organisms when no particular gear (tool) is involved. Furthermore, the same fishing gear can be used in different ways. A common way to classify fishing gears and methods is based on the principles of how the fish or other prey are captured and, to a lesser extent, on the gear construction.

The main categories of fishing gear as follows:

— Surrounding nets (including purse seines)

— Seine nets (including beach seines and Boat, Scottish/Danish seines)

— Trawl nets (including Bottom: Beam, Otter and Pair trawls, and Midwater trawls: Otter and Pair trawls)

— Dredges

— Lift nets

— Falling gears (including cast nets)

— Gillnets and entangling nets (including set and drifting gillnets; trammel nets)

— Traps (including pots, stow or bag nets, fixed traps)

— Hooks and lines (including handlines, pole and lines, set or drifting longlines, trolling lines)

— Grappling and wounding gears (including harpoons, spears, arrows, etc.)

— Stupefying devices

Fishing methods have continuously evolved throughout recorded history. Fishers are inventive and not afraid of trying new ideas. The opportunities for innovation have been especially good in recent decades with advances in fibre technology, mechanisation of gear handling, improved performances of vessels and motorisation, computer processing for gear design, navigation aids, fish detection to mention only a few technologies.

Whereas technological development of fishing gear and methods in the past was aimed to increase production, the present situation with many overfished stock, limited possibilities to expand fishing on underexploited resources and concerns about the environmental impact of fishing operation, gear development is now very much focussed on selective fishing and gears with less impact on the environment.

Gillnet

Mesh fishing nets are used in many different ways around the world. Larger fish are caught by their gill covers when they encounter the net and retreat, leading to the name "gillnet". Depending on how big the holes are, a gillnet can catch larger fish while allowing smaller fish to pass through. For example, Oregon fishermen use a specific mesh size that allows them to catch salmon while allowing smaller steelhead trout to escape.

Large nets for salmon and other fisheries are stored on a reel on the back of a boat which helps unwind the net into the water without tangling. This kind of net can be set at different depths between pairs of anchored buoys. Gillnets are generally retrieved after anywhere between 6 hours to a few days of soak time.

People in coastal communities often set small gillnets close to shore, while commercial operations set larger nets in deeper waters. Though a gillnet may be only 3m deep, it can stretch for 50 to 200m. Groups of nets can be tied together and extend from 300 to 3000m.

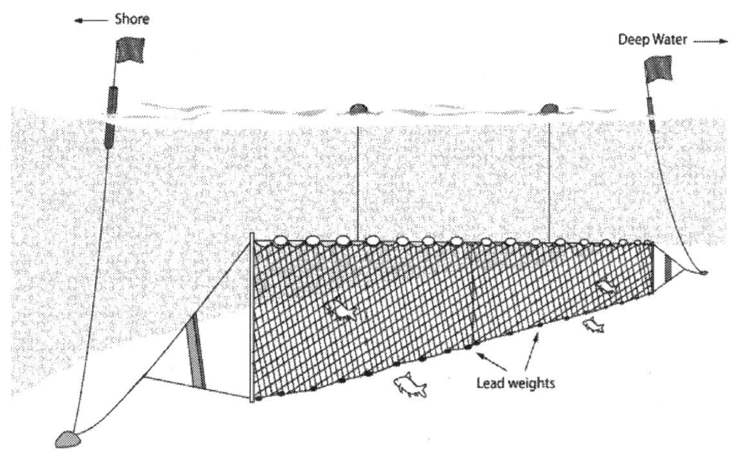

Gillnet

Set nets are attached to the bottom and held open with weights at depth and floats along the top. In contrast, drift nets are released by a boat and allowed to drift with the current. In some cases the boat remains attached and will drift with the net to avoid losing it.

A drift net that escapes will continue fishing, and without anyone to retrieve the catch, the net will continue to trap fish and marine life long after it has been abandoned. For this reason, large drift nets have been banned in the European Union, the Baltic Sea, and other areas of international waters.

Types of Gillnets

— *Set net, anchored or set gillnet* - nets are anchored in one place

— *Bottom gillnet* - nets are anchored along the seafloor

— *Midwater gillnet* - nets are anchored off the bottom, in water column

— *Drift net, drift gillnet* - nets without an anchor

— *Tangle net* - net with smaller mesh that catches fish by teeth rather than gills. Improved survival allows fishermen to keep hatchery salmon and release wild salmon.

— *Trammel net* - two or three layers of netting, including a middle layer with smaller mesh. Often used at or near the bottom.

Although static gears such as gill nets generally have less impact on the environment than mobile or towed gears they pose a particular problem for cetaceans (dolphins and porpoise). Methods to increase the 'dolphin-friendliness' of this fishing method include the attachment of acoustic devices or 'pingers' to the net to deter the animals; reducing the 'soak time',; i.e. the amount of time the net is left in the water;restrictions on the length of net used; and the introduction of closed areas to exclude fishermen from cetacean 'hot-spots'.

An EU Regulation (812/2004) came into force in January 2005 which lays down measures to: reduce incidental catches (by-catch) of cetaceans in fisheries through the mandatory introduction of acoustic devices (pingers) on vessels over 12m; monitoring of vessels (over 15m) in fisheries where by-catch of cetaceans has been implicated; phase out and eventually ban fishing with drift nets in the Baltic Sea. The first phase of the pinger requirements is to be implemented in certain North Sea fisheries by June 2005.

Gillnet Bycatch

Gillnets sometimes catch cetaceans along with the fish, including bottlenose dolphins, common dolphins, white-sided dolphins, dall porpoises, harbor porpoises, sperm whales and beaked whales, to list only a few. Bycatch in gillnets may be the single greatest threat to porpoise and dolphin populations worldwide. The vaquita, a tiny porpoise in Mexico's Gulf of California, is currently on the brink of extinction due to being caught in the gillnet fisheries for sharks and totoaba.

Extensive effort has been made to modify gillnets to either reflect or produce sounds in order to warn cetaceans away from the net, with varying success. Recent experimental studies of acoustic deterrents known as "pingers" show reduced bycatch for certain fisheries, though some scientists caution that pingers are not a panacea.

Other regulations sometimes limit the soak time and mesh size of gillnets in order to reduce bycatch. Brief time area closures in the Gulf of Maine yielded equivocal results. In the state of California (USA) and around the Banks Peninsula (New Zealand), specific areas have been closed to gillnet fishing seasonally or permanently in order to reduce marine mammal bycatch.

Longline Gear

Longline gear consists of a main line with many shorter lines of baited hooks trailing from it. Longliners use different sets of gear with varying hooks, bait, depth and timing to match the behavior of the species they are trying to catch. For example, swordfish sets are shallower than tuna sets, and typically use lightsticks in addition to bait. A combination of anchors and floats can be used to fix the lines at different depths. Swordfish are usually caught during the night, while tuna is caught during the day.

In general, longline gear targets large fish. Tuna caught by longline are bigger and more valuable than the smaller fish caught in great numbers by purse seine fisheries. In the Eastern Pacific, longline fishing peaks around the Marquesas Islands and near Peru.

The baited hooks of a longline attract more than just the target species, including pilot whales and other cetaceans which are sometimes caught as bycatch. A growing problem with longlines is the removal of hooked fish by sperm whales, killer whales, and false killer whales.

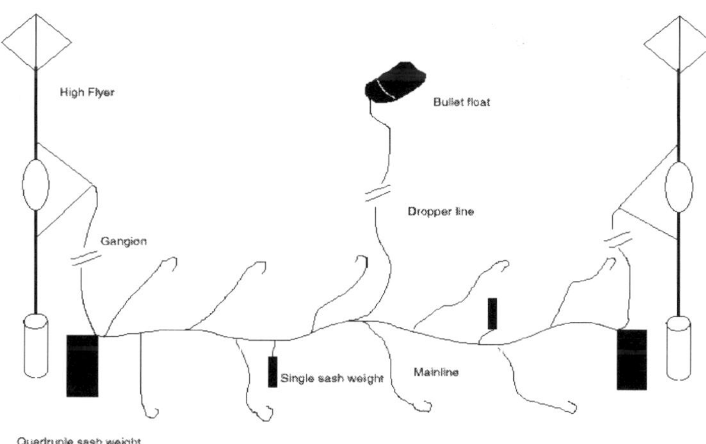

Bottom longline gear

Aside from the problem of hooking the cetaceans, some fishermen have resorted to rifles and even dynamite to protect their catch.

Types of Longline Gear

— *Bottom longline* - the main line is anchored along the seafloor

— *Pelagic longline* - the main line is anchored in the water off the bottom

— *Tuna set* - specialised hooks and lines used specifically to catch tuna

— *Swordfish set* - specialised hooks and lines used specifically to catch swordfish

Purse Seine

Purse seines are used to catch entire schools of fish. Fishers can identify schools by keeping watch for associated whitewater, dolphins, or seabirds at the surface, while scanning the depths with echo-sounders and sonar. Larger operations send spotter planes or helicopters to look for schools and notify the main fishing boat.

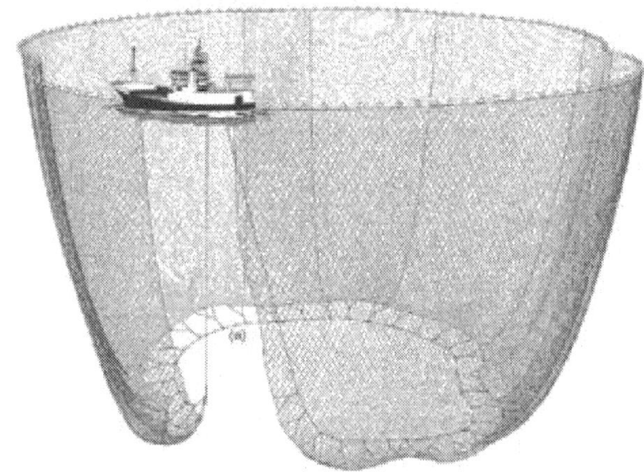

Purse Seine

When a school is found, the main vessel and assisting skiff encircle the fish with a net. The bottom is pulled closed around the fish with a drawstring known as a purse cable. The net is then pulled out of the water by the main vessel like a giant "purse" full of fish. More tuna is caught by purse seines than by any other fishery.

Purse seiners can also set their nets on passing debris (known as "logs") or fish aggregating devices (FADs), since fish are attracted to floating objects. At night, fish can be attracted to the net by shining light into the water. Different methods for targeting schools of tuna are known as "fishing on schools," "fishing on dolphins," or "fishing on logs," when whitewater, dolphins, or floating objects are used.

Fishing on dolphins yields larger yellowfin tuna than other purse seine sets, but can lead to incidental catch of dolphins. Helicopters may be used to spot dolphins, leading to their pursuit and capture. Specific methods for releasing dolphins from the net have been developed, allowing bycatch to be reduced in some regions.

Fishing on schools is sometimes less successful than other methods because the school of fish can disperse before the net is drawn. School sets are often coastal, and are especially common along the west coast of Baja California and the mouth of the Gulf of California (Mexico), the Gulf of Tehuantepec (Mexico) and the Gulf of Guayaquil (Ecuador).

Purse seines can also be set around any object floating at or near the surface, from tree trunks to sea turtles, whales, or discarded fishing gear. In coastal areas of Central America, fishing on tree trunks is concentrated where rivers carry debris from forested areas. Many other kinds of fish, sharks, and birds also associate with floating objects, leading to a wider range of bycatch than for other purse seine strategies.

Types of Purse Seines

— *American seiner* - power block hauls the net, less stable in rough weather

— *European seiner* - adapted to rough weather, special winch hauls the net

— *Block seine* - power block hauls the net, less stable in rough weather

— *Drum seine* - net is hauled onto a drum, requires fewer people to operate

— *Two boat purse seine* - two boats set the net quickly before fish disperse

— *One boat purse seine* - one boat with a skiff is slower but cheaper to run

Trawl

Trawl nets are dragged behind a boat, either along the seafloor or through the water. The cone-shaped net is held open with a beam or by heavy doors on either side of the mouth of the net. Pelagic or midwater trawls are pulled through the water column, though these nets may sometimes reach the bottom when full. Trawl fishing involves three basic steps. First the net shoots into the water from the boat. Then it is towed until the net is full. Finally, the net must be hauled back onboard, concentrating the catch in the very end of the net, known as the "cod end" or the "bag."

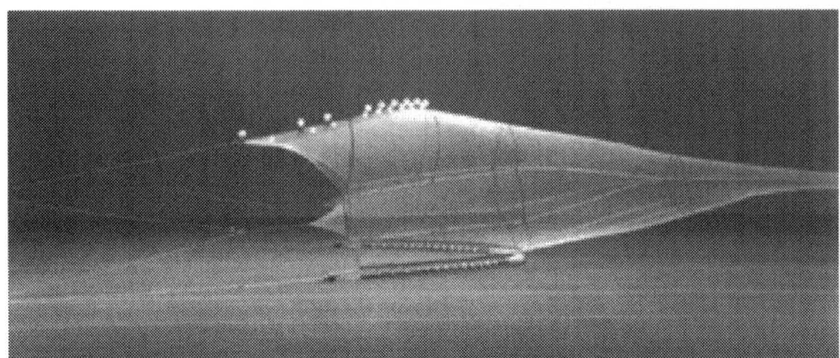

Trawl nets

Bottom trawls or draggers tow their nets across the seafloor. Historically trawlers have avoided rocky reefs, corals, and other structures on the seafloor to prevent tangling the nets. However, the recent development of rockhopper trawls with large wheels designed to roll over seafloor obstructions has allowed trawlers to fish in larger areas than ever before. Rows of rubber tires send the net bouncing along the seafloor in areas that were previously inaccessible. Both midwater and bottom trawls sometimes catch cetaceans. Animals caught as bycatch may be discarded after the catch is hauled onto the boat, but often do not survive.

Types of Trawl Gear

— *Bottom trawl* - trawl net is dragged along the seafloor

— *Beam trawl* - a beam holds open the net

— *Dragger* - trawl net is dragged along the seafloor

— *Midwater trawl* - trawl net is towed through the water, above the bottom

— *Outrigger trawl* - larger nets are towed and held open by outriggers extending from each side of the boat

— *Otter trawl* - net is held open by heavy doors known as "otter boards"

— *Pair trawl* - larger net is towed and held open between two boats

— *Pelagic trawl* - trawl net is towed through the water, above the bottom

— *Rockhopper trawl* - the net bounces along the seafloor with large wheels

In beam trawl the mouth or opening of the net is kept open by a beam which is mounted at each end on guides or skids which travel along the seabed.

The trawls are adapted and made more effective by attaching tickler chains (for sand or mud) or heavy chain matting (for rough, rocky ground) depending on the type of ground being fished. These drag along the seabed in front of the net, disturbing the fish in the path of the trawl, causing them to rise from the seabed into the oncoming net.

Beam trawl

Electrified ticklers, which are less damaging to the seabed, have been developed but used only experimentally. Work is also being carried out to investigate whether square mesh panels fitted in the 'belly' or lower panel of the net can reduce the impact of beam trawling on communities living on or in the seabed. Modern beam trawls range in size from 4 to 12 m (weighing up to 7.5 tonnes in air) beam length, depending on the size and power of the operating vessel.

The demersal or bottom trawl is a large, usually cone-shaped net, which is towed across the seabed. The forward part of the net – the 'wings' – is kept open laterally by otter boards or doors. Fish are herded between the boards and along the spreader wires or sweeps, into the mouth of the trawl where they swim until exhausted. They then drift back through the funnel of the net, along the extension or lengthening piece and into the cod-end, where they are retained.

The selectivity of trawl fisheries may be increased by the use of devices known as separator trawls. Separator trawls exploit behavioural differences between fish species and can be used, for example, to segregate cod and plaice into the lower compartment of the net, whilst haddock are taken in the upper part. The mesh size for the two compartments can be altered according to the size of the adult fish being targeted. Insertion of square mesh panels also improves selectivity of the net because square meshes, unlike the traditional diamond shape meshes, do not close when the net is towed. Discarding of immature fish may also be reduced by increasing the basic mesh size in fishing nets. Sorting grids are compulsorily fitted in nets in some prawn and shrimp fisheries to reduce bycatch of unwanted or non-target species, including small prawns and shrimp.

Depending on the depth of water fished and the way in which the gear is constructed and rigged, trawling may be used to catch different species. Trawls can be towed by one vessel using otter boards, as in bottom-trawling, or by two vessels, each towing one warp, as in pair-trawling. Or more than one trawl can be towed simultaneously as in multi-rig trawling.

Multi-rigs

Multi-rigs are used widely for the capture of panaeid shrimps in tropical waters and more recently for Nephrops (langoustines or Dublin Bay prawns) and deep-water prawns in temperate waters. The speed at which the net is towed is important, varying with the swimming speed of the target species from about 1.5 to 5 knots for fast swimming fish.

Dive-caught

Free diving (using mask and snorkel) or scuba diving is a traditional method of collecting lobster, abalone, seaweed, sponges and reef dwelling fish (groupers and snappers) for example. In deeper waters helmet diving systems using air pumped from the surface are used. Species, including high value species such as geoduck (giant clam), urchins, sea cucumber, lobster and scallops are now widely harvested by divers. Hand-collection by divers is potentially one of the most species selective and least damaging fishing method, provided harvesting is carried out responsibly.

Dredging

Dredging is used for harvesting bivalve molluscs such as oysters, clams and scallops from the seabed. A dredge is a metal framed basket with a bottom of connected iron rings or wire netting called a chain belly. The lower edge of the frame has a raking bar, with or without teeth, depending upon the species targeted. The catch is lifted off the seabed or out of the sea by the raking (or teeth) bar and passes back into the basket or bag. Depending on the size of the boat and the depth of water fished the number of dredges or 'bags' may vary from a single dredge towed behind the vessel to from 5 to

10 or more dredges per side. Dredges are generally attached to a towing bar and one is operated from each side of the vessel simultaneously.

Drift net

A gill net that is allowed to drift with prevailing currents. Drift nets are not set or fixed in any way, are in fact 'mobile', and they are allowed to drift with the prevailing currents. Drift nets are used on the high seas for the capture of a wide range of fish including tuna, squid and shark, and off north-east England for salmon. Despite a global moratorium on large-scale drift nets (nets exceeding 2.5 kms in length), introduced in 1992, problems still exist. For example, drift net fisheries in the Mediterranean for swordfish and albacore tuna pose a particular threat to striped dolphins. An EU-wide ban on all drift nets was introduced from January 2002.

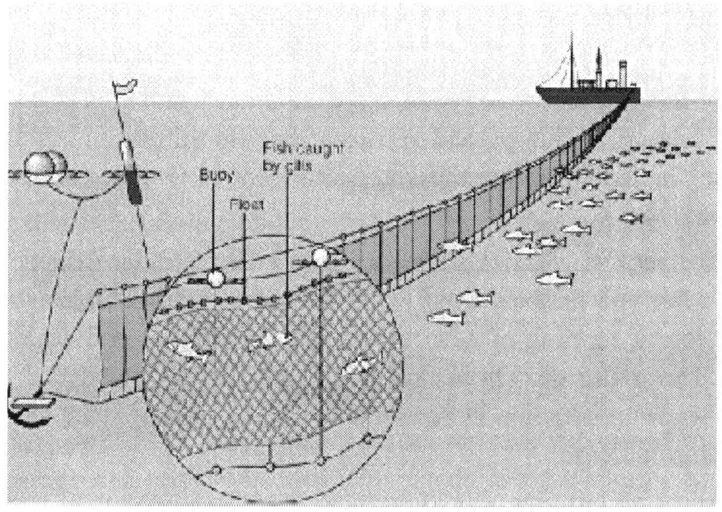

Drift-net in use

The ban applies to fisheries such as tuna, shark and swordfish in all EU waters except the Baltic, and to all EU vessels on the high seas. EU fishermen are, however, considering challenging the ban if 'pingers' are found to be successful in deterring marine mammals from entanglement and subsequent drowning in nets.

Fish Attraction Devices (FADs)

Various species of fish often congregate or associate with other living creatures (e.g. tuna associate with dolphins and whale sharks) or objects

pump to suck bottom sediments on board ship where bivalves are screened out and the spoil discharged back to sea. Impacts associated with this type of fishing are removal of local populations of the target species, removal and disturbance of sediment with consequences for other species living there, and creation of spoil plumes and siltation.

Industrial fishing

Most fishing methods target fish for direct human consumption. Fisheries targeting species for reduction purposes i.e. the manufacture of fish oil and meal, are referred to as industrial fisheries. Fish meal and oil is produced almost exclusively from small, pelagic species, for which there is little or no demand for direct human consumption. The methods of capture are purse-seining and trawling with small mesh nets in the range of 16-32 mm. Important industrial fisheries in South America include the Chilean jack mackerel fishery and the Peruvian fishery for anchoveta.

Industrial species in the North Sea and North-East Atlantic include: sandeel, sprat, capelin, blue whiting, Norway pout and horse mackerel. Fish oil is used in a range of products including margarine and biscuits. Fish meal and oil has more widespread use, however, in the manufacture of pelleted feedstuffs for intensively farmed poultry, pigs and, not least, aquaculture.

One of the main impacts associated with industrial fishing is the removal of large quantities of species from the base of the food chain. For example the sandeel fishery in the North Sea, the largest single-species fishery in the area accounting for over 50% by weight of total fish landings, has been implicated in the decline of breeding success in seabirds such as kittiwakes, and reducing food availability for marine mammals and other commercial fish species such as cod and haddock.

Long-lining

Long-lining is one of the most fuel-efficient catching methods. This method is used to capture both demersal and pelagic fishes including swordfish and tuna. It involves setting out a length of line, possibly as much as 50-100 km long, to which short lengths of line, or snoods, carrying baited hooks are attached at intervals. The lines may be set vertically in the water column, or horizontally along the bottom. The size of fish and the species caught is determined by hook size and the type of bait used.

Although a selective method of catching fish, long-lining poses one of the greatest threats to seabirds. Species such as albatross, petrels, shearwaters and fulmars scavenge on baited hooks, get hooked, are dragged underwater and drowned.The problem occurs whilst the baited hooks are on or near the surface i.e. before the hook sinks. Commonly the bait used is squid, the principal prey of many seabird species. Most globally threatened species, including the majestic wandering albatross, live in the Southern Ocean.

A range of practical measures have been developed to help prevent seabirds being hooked and drowned on longlines. These include bird-scaring streamers that flap and scare birds away, setting lines at night when most albatross do not feed and weighting the line so it sinks quickly, bird scaring water cannons and setting the line nearer the water surface rather than over the side of the boat,thus minimising the lenght of time the bait is visible/ available. Any of these measures will contribute to reducing seabird by-catch. Ask your supplier if the longline caught fish you buy has been caught using "seabird-friendly" methods.

Pelagic trawl

When trawling takes place in the water column or in mid-water between the seabed and the surface, it is referred to as mid-water or pelagic trawling. Pelagic trawls target fish swimming, usually in shoals, in the water column i.e. pelagic species. These include seabass, mackerel, Alaska pollack, redfish, herring and pilchards for example. Their effectiveness relies on traversing a considerable volume of water, and consequently nets are larger than bottom trawls and require a large vertical and horizontal mouth opening to provide net stability and capture large shoals of fish. The length of time the net is towed through the water is shorter than in bottom trawling in order to capture the shoals of fish the net passes through. To handle the large amounts of fish, pumps are used to transfer the catch from the cod-end to the boat.

In mid-water pair trawling the otter boards are replaced, and the mouth of the net kept open, by a pair of trawlers. This enables vast nets, often ¼ mile wide and ½ mile long, to be towed through the water column to capture the fish.

Pelagic trawling affects marine mammals as they are caught accidentally when feeding on the same fish being targeted by fishermen; being unable to surface for air they eventually drown. Capture of marine mammals in fishing nets represents a very significant welfare problem. Animals can remain conscious for some time while struggling in the net, causing suffering and injuries such as lacerations and broken teeth and bones, before dying of suffocation. In response to the continuing slaughter of dolphins in pelagic trawl nets targeting seabass and other species, the UK Government is developing a marine mammal escape device which, if successful, will reduce the number of dolphin casualties in these fisheries.

Pole and line

Pole and line fishing (also known as bait boat fishing) is used to catch naturally schooling fish which can be attracted to the surface. It is particularly effective for tunas (skipjack and albacore). The method almost always involves the use of live bait (anchovies, sardines etc.) which is thrown over board to attract the target species near the boat (chumming). Poles and lines with barbless hooks are then used to hook the fish and bring them on board. Hydraulically operated rods or automatic angling machines may be used on larger pole and line vessels.

Pots (or creels)

Pots (or creels) are small baited traps which can be set out and retrieved by the operating vessel. They are widely used on continental shelves in all parts of the world for the capture of many species of crustaceans and fish, together with octopus and shellfish such as whelks. Potting is a highly selective method of fishing, since the catch is brought up alive, and sorting takes place immediately, allowing unwanted animals to be returned to the sea, making the method potentially sustainable.

Pots (or creels)

However, in Britain, fishing effort in the potting sector is high, with currently no restrictions on the number and type of pot used or the amount of shellfish taken. Pots used to be constructed from 'withy' or willow, but are now constructed from plastic-coated or galvanised wire with nylon netting. This makes them virtually indestructible.

Modern pots or 'parlour pots' are also more complex and fitted with 'pot-locks', making escape impossible for the crab or lobster entering it. These factors combined with mechanical hauling allow fishermen to haul more pots and to leave them on the seabed for longer, thus increasing efficiency and fishing capacity.

Seine netting

This is a bottom fishing method and is of particular importance in the harvesting of demersal or ground fish including cod, haddock and hake and flat-fish species such as plaice and flounder. The fish are surrounded by warps (rope) laid out on the seabed with a trawl shaped net at mid-length. As the warps are hauled in, the fish are herded into the path of the net and caught. Effectiveness is increased on soft sediment by the sand or mud cloud resulting from the warps' movement across the seabed.

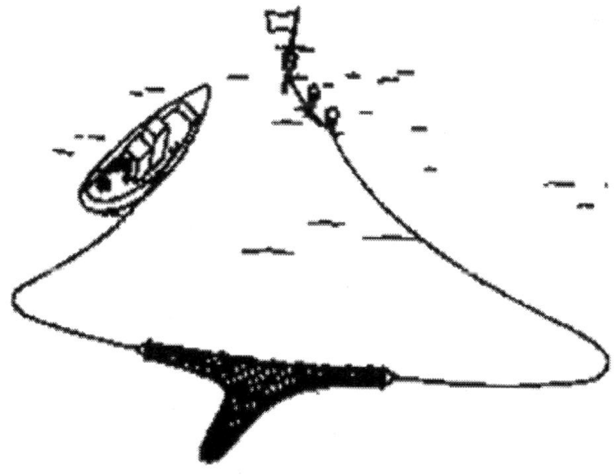

Seine netting

This method of fishing is less fuel-intensive than trawling and produces a high quality catch, as the fish are not bumped along the bottom as with trawling.

Selectivity of Gear

Selective fishing refers to a fishing method's ability to target and capture organisms by size and species during the fishing operation allowing non-targets to be released unharmed.

In a fishing area, a range of species and sizes normally occur together, both fish and non-fish. When encountered by a fishing gear, they will be captured at different rates, depending on the gear design and its mode of operation. Various gears are known to be more selective in this regard than others. A purse seine has a poor selective performance, as it will catch most of the surrounded individuals. No gear is known to be one hundred percent selective for a given species or size range of individuals.

Most gears and methods fall into a group that has some selective performance whose ability to select targets can be altered through modification of design and operation methods. The catch in many types of fisheries thus consists of a mixture of targets and non-targets. What does or does not comprise targets depends to a large extent on the market and whether there are regulations in place prohibiting capture of species or certain sizes of organisms.

Non-targets are often synonymous with bycatch, a concept defined differently by various people. A generic definition often used is that bycatch refers to the capture of any species, size of species, or sex of species that is not the primary target(s) of a fishing activity. Bycatch, as a choice of the fisher, may be retained or discarded, alive or dead. The significance of the discard depends on biological, ecological, or socio-economic factors. The nature of discards can possibly be classified into four categories:

1) population status,
2) socio-economic,
3) ecological and
4) public concerns.

These factors most frequently form the basis or regulatory actions taken by management agencies. Total global discard is difficult to estimate, but one assessment came up with a level of 27 million tonnes for the period 1988 through 1990. A 1996 FAO Technical Consultation in Tokyo concluded, however, that the 27 million tonnes might be an overestimate, elaborating different reasons.

Besides bycatch of fish and fish-like species, other animals might incidentally be captured with fishing gears. This includes, among others, various species of whales, turtles and seabirds . Although bycatch of such organisms seldom is a threat to their population size, public concerns make it necessary to reduce such bycatch.

Improved selectivity can be achieved in different ways, by modifying the gear design and/or operation and by using alternative fishing gears. In trawls and gillnets, mesh size is a well-known measure to regulate the size of captured organisms. For mobile gears, like trawls and seines, improved selectivity can also be achieved by using square meshes in the codends and by inserting filtering grids in front of the codend. Successful separation of targets and non-targets species can also be achieved by using grid devices.

he principles for such selectivity are detailed in the referenced documents. Successful technical solutions have also been found to reduce capture of non-fish species like mammals, turtles and seabirds. The capture of dolphins in the purse seine fishery for tuna has been reduced to an insignificant level by using a combination of technical changes, rescue techniques, education of fishers and management actions.

Bycatch of turtles in the tropical shrimp fishery can be avoided by using a turtle excluder device (TED) is a rigid or soft (netting) structure inserted in the aft part of a trawl, in front of the codend proper.The incidental capture of seabirds in longline fisheries can be significantly reduced by underwater setting of the line, night setting and scaring the seabirds away from the baited hooks.

At international level the most prominent action to reduce bycatch during fishing is the ban on large size driftnetting on the high seas (UN General Assembly Resolution 1990). The FAO Code of Conduct for Responsible Fisheries of 1995 is a major international document addressing the problem with selective fishing. The most recent international contribution is the International Plan of Action for Reducing Incidental Catch of Seabirds in Longline Fisheries adopted by the FAO Committee on Fisheries in 1999.

FAO has recently developed a project (with GEF funding), to enhance the capability of developing countries to reduce environmental impact of tropical shrimp trawling through the introduction of bycatch reduction technologies and change of management. At national and regional levels regulations, minimizing unwanted bycatch and discards are being introduced.

These include minimum mesh sizes, mandatory use of selective devices like TEDs and sorting grids, areas closed for fishing, etc. Many countries are putting a great deal of effort into research and development aimed at improving the selectivity of fishing gear and methods that presently capture much of the unwanted catch.

Exclusive Economic Zones (EEZ)

The Exclusive Economic Zone (EEZ) is the area adjacent to a coastal state which encompasses all waters between:

a) the seaward boundary of that state,

b) a line on which each point is 200 nautical miles (370.40 km) from the baseline from which the territorial sea of the coastal state is measured (except when other international boundaries need to be accommodated), and

c) the maritime boundaries agreed between that state and the neighbouring states.

In simpler terms, it is a zone under national jurisdiction (up to 200-nautical miles wide) outlined in the provisions of the 1982 United Nations Convention of the Law of the Sea, within which the coastal State has the right to explore and exploit, and the responsibility to conserve and manage, living and non-living resources.

The EEZ resources may be entirely contained within the EEZ limits and be managed solely by the coastal state in conformity with the provisions of the The United Nations Convention on the Law of the Sea. Some of these resources may, however, extend beyond EEZ boundaries, into the neighbouring exclusive economic zone (shared stocks) requiring sharing and management agreements with neighbouring coastal states. Some may also extend into, or migrate through, the high seas (straddling stocks) and as such need to be managed in cooperation with other countries, usually under the aegis of a regional fishery management organisation.

The basic rules for such agreements are provided by the 1995 UN Agreement for the Implementation of the Provisions of the United Nations Convention of the Law of the Sea of 10 December 1982 Relating to the Conservation and Management of Straddling Fish Stocks and Highly Migratory Fish Stocks (often referred to as the UN Fish Stocks Agreement) which is not yet in force. The FAO Code of Conduct for Responsible

Fisheries, adopted in 1995, is the main guide to the practical implementation of all the relevant instruments needed for fisheries governance.

The challenge is to manage fisheries in such a way as to ensure the optimum and sustainable use of resources as well as economic efficiency and widespread social benefits. Governments have the responsibility for fisheries management. They may delegate such responsibility to specialised agencies and should promote effective participation, including in decision-making, of those operating in the fisheries sector or also using the aquatic resources and their environment for different purposes.

More than ninety percent of the global fish catch is taken within zones that are under national jurisdiction. It follows, therefore, that these are the areas where most fisheries management problems occur.

Such problems are not new: for more than 50 years, at least since the London Conference on Overfishing (1946), recognition has been given to the need for governments to be aware of the state of their fisheries, to adopt policies aimed at preventing the depletion of resources and a wastage of fisheries inputs and, increasingly, to assist in the recovery and rehabilitation of over-fished stocks. With about 25 percent of the main commercial stocks for which data is available being overfished and 50 percent more being fully exploited, the need for better governance is clear.

In the 1980s it was widely anticipated that fisheries governance (and state of resources) would improve substantially in parallel with the establishment of extended national jurisdiction. Subsequent experience has shown that, even under the most favourable circumstances, achieving good governance and resource sustainability is a protracted and difficult process.

Those governments that now have soundly managed fisheries in their EEZs generally owe this achievement to 20 to 40 years of continuous effort and adjustment. In many countries, governance has continued to languish for a variety of reasons, including inadequate institutions (including unclear users' rights), a scarcity of the human and financial resources required to devise and implement management programmes; a lack of understanding, by both governments and fisheries participants, of the potential benefits that good management can generate; and the reluctance of governments to make unpopular decisions.

Developing countries have become major participants in the global fishery sector but most of them require improvements in their fisheries governance and financial as well as technical assistance. They require the scientific, technical and administrative capabilities necessary to formulate and implement appropriate fisheries management plans and to assess their outcomes and to take any necessary follow-up action. Unfortunately, the countries with the poorest fisheries governance are often those whose populations face more pressing, fundamental problems such as war, civil disturbances and natural disasters.A fundamental policy consideration is the strengthening of fisheries institutions so that they have the capacity to provide independent technical advice and guidance throughout the sector.

The trend in the governance of fisheries is for management functions to devolve progressively from government, without detracting from its stewardship role, to include the direct involvement of fisheries participants, the conferring of user rights and the financing of governance from within the sector. At the same time, it must be accepted that improved benefits will not be an immediate outcome from better governance. The structural adjustments that are required in many fisheries will take a long time to become effective.

Destructive Fishing Practices

All fishing methods have an impact on the target resource and may also affect non-target species. Many of them also have an impact on the wider aquatic environment. But the "normal" effect of exploitation should not be confused with "destruction" which, according to the dictionary implies "to reduce to a useless form, to spoil completely; to put out of existence; to obliterate, wipe out, annihilate, demolish, devastate, tear down, raze". The term "destructive fishing" has been generally used by scientists and environmental NGOs for a wide range of situations spanning from classical overfishing (excessive use) to outright destruction of the resource and its environment with explosives (wrong methods). Indeed, many fishing gears could be considered "destructive" if used in the wrong environment.

In a fisheries management perspective, it is however important to distinguish among:

— overuse or overfishing, to be corrected mainly through fishing capacity reduction;

— non-selectivity, which in most cases can be corrected through improved gear technology, closed seasons, or closed areas, etc.;

— destructive fishing, or destructive use, e.g. when a gear is used in the wrong habitat, such as bottom trawls in seagrass beds where its impact is unacceptable, and where the gear can be prohibited; and

— destructive methods, the impact of which are so indiscriminate and/or irreversible that they are universally considered "destructive" in whatever circumstances.

Among the latter, the use of poisons and explosives is prominently cited in the literature and the muroami fishing practice is often referred to (in Southeast Asia). The economic losses for the local and wider society resulting from such damage far outweigh the short-term individual gains made by the users of these destructive methods.

Use of Poisons

The use of poisons is widespread, in some regions in both fresh and marine waters—especially in coral reefs and coastal lagoon fisheries. In many places the use of poison is a traditional practice but the effects have been exacerbated by the use of pesticides to replace poisons of vegetable origin. As fish become scarcer through overfishing or in order to catch rare, small and precious aquarium fish, local fishers often resort to using poisons such as cyanide or pesticides. Fishers can easily obtain inexpensive cyanide used in the jewellery industry and gold mining. Pesticides are readily available to farmers, which are often also part-time fishers. Techniques used vary across regions/localities. They are effective at killing or stunning, indiscriminately, the fish, which are then collected by divers, or through netting and seining. The poisons kill also other organisms from the ecosystem, including the coral reef-building organisms.

Fishing with Explosives

Fishing with explosives, also known as blast fishing, has probably been in existence for centuries and is apparently spreading. Explosions can produce fairly large craters, devastating 10-20 square meters of bottom. In coral reefs, recolonisation of damaged habitats is very slow and complete recovery may take several decades. The explosion kills both the target fish and the accompanying fauna being indiscriminate in size or species. In many instances, errors of manipulation have lead to injuries and death of humans.

Explosives and the raw materials for preparing explosives such as fertilisers and sugar are cheap and easily available. Commercial explosives are often obtained from mining or construction activities. In many areas, fishers only need to extract the explosive charges from munitions left over from on-going or past armed conflicts. In other areas, fishers can access army munitions through illegal channels. Explosives can have very serious consequences for the resources, the environment, and unfortunately sometimes also for the users themselves.

Muroami Fishing Technique

The Muroami fishing technique, employed on coral reefs in Southeast Asia, uses an encircling net together with pounding devices. These devices usually comprise large stones fitted on ropes that are pounded onto the coral reefs. They can also consist of large heavy blocks of cement that are suspended above the sea by a crane fitted to the vessel. The pounding devices are repeatedly and violently lowered into the area encircled by the net, literally smashing the coral in that area into small fragments in order to scare the fish out of their coral refuges. The "crushing" effect of the pounding process on the coral heads has been described as having longlasting and practically totally destructive effects.

Inadequate Practices

Some standard fishing gears could be used in a way which damages the resource and/or the environment, to such an extent that they could be (and have been) considered as "destructive" fishing practices. These may include, inter alia, the following practices that can be properly regulated and controlled:

— beach seining, because of the large proportion of juveniles yielded and often discarded;

— bottom trawling because of its impact on the bottom and on bottom-dwelling animals and benthos such as sea stars, urchins, clams, etc. In addition, when improperly used in the wrong environment e.g. on coral reef areas or coastal seagrass beds (despite of being generally prohibited in such areas) it may have very longlasting effects on habitat;

— large-scale pelagic driftnets because of their ability to ensnare large marine animals such as mammals, sharks, turtles, and a number of vulnerable species;

— fishing with respiratory assistance using scuba equipment or compressors, because their effectiveness is judged excessive;

— anchoring of vessels in reef areas (particularly the Staghorn coral) which breaks them, seriously damaging the whole reef sustainability.

The techniques universally recognised as "destructive" can only be explicitly banned and their use severely punished. Most bad ways of using fishing gear have been identified and are prohibited in national legislation. However, experience shows that this is not sufficient because in stressing economic situations, the incentives to misbehave might be very high.

For the poor, poisons and explosives are among the cheapest method available, requiring little capital investment. In some instances, it is the only affordable access to a vital resource. Often, however, the replacement of poison and explosives by other types of fishing methods is within the financial means and reach of those who currently use these destructive fishing practices.

For the poorest, however, it will be difficult to suppress these bad fishing habits without improving their economic condition. In these instances, the long-term solution to the use of destructive methods is in improving livelihoods or providing affordable alternatives. The Muroami technique does not seem to be associated specifically with poverty. It reflects a total disregard for sustainability and as such should be banned. The destructive modes of use of other more standard gears can also be "promoted" by chronic overfishing, in a very damaging vicious circle. Proper practices should be regulated through protection of particularly important areas, or habitats (seagrass beds, coral reefs) as well as closed seasons (particularly to protect juveniles or sensitive species).

Eliminating destructive methods requires finding solutions to a number of issues affecting the fisheries sector such as overfishing (an example of "the hen and the egg" dilemma), inequitable resource allocations, food insecurity and poverty. In most cases, there is a need to improve stewardship over the resources through better monitoring and control, educating fishers about the destructive nature of their practices, and the establishment of explicit forms of use rights for local communities together with decentralised management responsibilities as well as no-take areas. There is evidence that the allocation of use rights will create a strong incentive for communities to use responsible fishing practices and persecute and punish trespassers.

Awareness about these methods and their effect is growing. All national fishery laws generally prohibit poisons and explosives but enforcement is inadequate. Large-scale pelagic driftnets have been banned by the United Nations General Assembly. The use of beach seines is prohibited in many countries and areas. The coastal areas with coral reefs or coastal seagrass beds are usually prohibited to trawls.

It should be recognised, however, that enforcement is often imperfect, particularly when offshore resources are overexploited. Programmes have been implemented to educate small-scale fishing communities of the damage caused by explosives and cyanides. The devolution of fishing rights and responsibilities to coastal communities (e.g. in Philippines) and their greater involvement with other stakeholders and NGOs in the management of resources, accompanied by programmes to raise awareness of the long-term damage of such fishing techniques, have contributed to the promotion of local social pressure and enforcement to control destructive fishing.

References

Bell, J. 1999. "Aquaculture: a development opportunity for Pacific islands" *Development Bulletin Information*. Paper 19:1-9.

Bonell A.D. 1994. *Quality assurance in seafood processing; A practical guide*. Chapman & Hall, New York, London.

FAO. 1995. Code of conduct for responsible fisheries. Rome, FAO, 41 pp.

Jakobsen M., Lillie A. 1992. "Quality systems for the fish industry". In *Quality assurance in the fish industry*. Eds: H.H. Huss, M. Jakobsen, J. Liston.

Sinha, V.R.P. 1996. "Shrimp Culture, Trade and the Environment in India". *INFOFISH*, Kualalumpur (Mimeo), 31p.

4

Administration of Capture Fisheries

Growing concerns of many developing countries include their current inability to control fisheries in their Economic Zones, the apparent high cost to do so, and the failure of some of the developed country fisheries which were formerly used as examples of fisheries management. The latter failures appear to have resulted from difficulties in implementing appropriate fisheries management strategies. In many countries there are no established systems for monitoring, control and surveillance of maritime zones to conserve their marine fisheries and associated habitats. Such countries need assistance in addressing this situation, to reap the potential benefits for their citizens from appropriate fisheries and marine resource management.

Fisheries Monitoring, Control and Surveillance (MCS)

Monitoring, control and surveillance (MCS) have often been seen as selfexplanatory and not requiring much consideration, as it is perceived as "simply the policing of the various maritime zones controlled by the State" - is it not? This technical document attempts to clarify this erroneous view of MCS and demonstrate how monitoring, control and surveillance are important to the implementation of any oceans related policy, in particular for fisheries management.

There are many influencing factors and global initiatives which have brought the subject of monitoring, control and surveillance to the fore in the international and national fora. The coming into force of the 1982 United Nation Convention on the Law of the Sea in November of 1994 has recognized the benefits of the Convention and once again raised the profile of the obligations States have with respect to the assessment of their fishing stocks, the allocation of the surplus to national needs to third parties and the further obligation to conserve their fisheries, including the fisheries habitat.

Further, the United Nations Conference on Environment and Development (UNCED) in Brazil in 1992 and the resultant Chapter 17 of Agenda 21 emphasized the urgent requirement to preserve our marine/fisheries environment on a global basis. The initiatives of several agencies on controlling pollution at sea; the United Nations Environment Programme's Global Environmental Facility Fund and Regional Seas Project and port State control initiatives for the merchant fleet, are all contributing towards global standards to protect our marine resources. The FAO flagging initiative to assist in the control of fishing vessels on the high seas, the potential of extending port State control and international co-operation with respect to fishing vessels, the efforts at responding to the need to establish standards for safety-at-sea for fishing vessels, and the most recent efforts at establishing standards for the control of fishing on the high seas of highly imgratory species and straddling stocks, are all symbolic of the growing global trend and concern regarding fisheries resources and their habitat. Finally, there is also an initiative to establish a "code of conduct", one might call it, for responsible fishing practices. This is all applicable to fisheries and the fisheries habitat.

As these Conventions come into force and the initiatives gain credibility, so too do the obligations of participating States to implement these agreements to which they are signatories. Many States have, and are, reaping the benefits of these agreements as internationally respected legal instruments, but now as they come into force, a case in point being the Convention, there is a realization that the implementation of these agreements carry considerable obligations for each State. Monitoring, control and surveillance (MCS), especially for the fisheries sector, is the implementing tool to meet these obligations. MCS, which has often been thought of as a luxury for developed States, has now become an obligation for all States to work together to conserve the marine resources and their environment. As an initial step, many countries are now focusing on developing and establishing policies and strategies to meet their obligations under the 1982 United Nations Convention on the Law of the Sea.

The ideal situation, and a logical approach to implement the terms of the Convention, would be to develop an oceans policy which would establish government priorities and the strategy for the conservation and sustainable use of all marine resources within the zone. From this oceans policy would flow the integrated oceans planning and management framework under which fisheries management would be developed. Most countries do not have the luxury of this longer term development process and have chosen to track in parallel these two initiatives; oceans policy development on the one hand, and fisheries management, including MCS strategies, on the other. Consequently, although countries remain cognizant of the fact that fisheries management must link to overall oceans policy when it is ultimately established, the development of the MCS systems required to implement fisheries management plans must go forward to address the more immediate need to conserve fish stocks and their habit. The parallel tracking of the development of oceans policy and fisheries MCS initiatives will impact on the latter as they are implemented, most probably by forcing the multiple tasking of fisheries MCS resources to address other related, maritime priorities. This impact will also depend on the focus of political will and the priority of fisheries in the overall oceans strategy. These points will obviously serve as an incentive, in the first instance, for countries to develop the most cost effective and appropriate MCS system to meet fisheries requirements, with the flexibility to assume other, secondary, fisheries-related tasking, such as pollution and environmental monitoring, as delegated by the government.

Fisheries management and the resultant MCS activities will depend on the decisions of government as to the level of its involvement in the fishing industry. Government can assume a very intrusive role, such that it actually runs the industry, controls potential income of fishers, and micro-regulates the harvesting sector. Alternatively, it can maintain a less intrusive role, whereby the fishing industry accepts its responsibilities and role within the framework of general government conservation principles and legislation. The South Pacific Forum Fisheries Agency has adopted this latter style of fisheries management. In either case, there are several perceptions regarding the fishers and MCS activities which have negatively coloured fisheries management, and fisheries MCS itself. There is the idea that many of the problems with fisheries in each country stem from the illegal fishing activites of foreign fishing fleets. Although foreign fishing fleets have had, and continue to have, an impact on fisheries conservation efforts in the majority of cases, the greater impact on fisheries in each country usually stems from its own domestic fishing industry in the coastal and nearshore fishing zones. Fisheries MCS activities focused solely on the control of foreign fishing efforts will therefore undoubtedly be less than successful as a key component in the conservation of the resources, if the domestic fishing effort is not also being addressed.

Another erroneous perception with respect to MCS is that it is a simple exercise of "activating the military might of the State to arrest the foreign poachers". Activating the military infrastructure for MCS is an initiative many countries have taken in the past and found to be prohibitively expensive and politically sensitive. Further, MCS activities are not solely enforcement operations, but also include the collection of data, and quality control of that data, for input into the stock assessment, social and economic, and enforcement exercises that comprise the components of fisheries management as well as safety-at-sea. This is not the normal use of military enforcement machinery and is neither applicable, nor appropriate, for domestic fishers.

The expense of MCS activities is usually the primary concern of any government in designing and implementing a system of controls. Cost effectiveness and efficiency are key to successful MCS operations. The surveillance, or enforcement, aspect of fisheries MCS has proven to be most cost effective and responds best to the political sensitivity of international fisheries incidents when it remains a civil action, without the potential involvement of the military might of the nation. Further, on the matter of expense, military machinery costs much more to build and operate than equivalent civilian equipment for a civil action. As an example, this can be verified by those who have attempted to place a civilian radio system on a piece of military equipment. There are significant additional costs associated with such an initiative in order to make the equipment meet the appropriate standards for military acceptance. Another cost consideration is whether a country requires a large armour-plated, heavily crewed vessel with associated high firepower and costs to carry out fisheries patrols and exercise civil police duties of data gathering and enforcement. The latter function can be exercised for all fisheries in a fully satisfactory manner with significant cost savings using a civilian vessel with fewer crew, less armour and firepower and hence, lower fuel consumption and general costs.

The basic needs for fisheries MCS normally include vessels that can remain at sea with the fishing fleets, appropriately equipped aircraft for cost-effective rapid surveillance and information collection and, an adequate coastal support infrastructure for verification of landings and the monitoring of the port trade of fish products.

Effectiveness of operations can be enhanced considerably if only one ministry is responsible, or has the lead role and authority, for the implementation of MCS activities, instead of two or three. This significantly reduces the lines of communications for the command and control, to use military terminology, of the monitoring and surveillance components for MCS activities, making them more efficient and responsive in a timely manner. Further, bureaucracy is usually significantly reduced using vessels under one ministerial control, as opposed to a combined civilian and military operation. These points are not to suggest that a country should completely obviate the use of military equipment as an option where no other options are available, but the military option is one which has fairly considerable consequences in terms of cost and control.

It has also been found through the experience of both developed and developing countries that MCS operations can be expedited with reduced costs and in an effective manner through co-operation with neighbouring countries on bilateral, sub-regional or regional initiatives. Examples of this include initiatives of the South Pacific Forum Fisheries Agency (FFA) and the Organization of Eastern Caribbean States (OECS) and initiatives being established in West Africa. Considerable cost savings can also be realized through the appropriate use of licenses as a control tool, as a source of basic information for management planning, and as an alternative to a free access fishing scheme while minimizing the cost to the resource owners, the taxpayers. As a practical example, licenses can require all transhipment of fish to occur in port, where they can be monitored much more accurately and safely, thereby contributing to MCS in a cost-effective manner. Fishing vessels could be encouraged to use ports by a reduction in administrative and bureaucratic requirements compared to regular transport and shipping vessels. This could also enhance the opportunity of input into port State control initiatives which States may wish to initiate in the future.

These types of cost saving strategies contribute to the implementation of "no [minimal] force" MCS strategies to the benefit of the coastal State.

A point made for years by fisheries officers is the issue of creating unenforceable legislation. Unenforceable legislation, or that which is not understood nor acceptable to the fishers, rapidly destroys the credibility and support for a government in its efforts to conserve its fisheries resources. Such legislation usually results in active subversion of its intended benefits by the fishers and the fishing industry. One example is the prohibition of catching a certain species of fish. Unless the individual is seen in the act, it is almost impossible to enforce. The prohibition would be best stated as it being against the law to be found in possession of this certain species of fish above a certain acceptable bycatch or tolerated level.

A somewhat controversial example of fisheries management systems which are difficult to enforce in developing countries is the legislation creating individual transferable quotas (ITQs). Although this system has all the positive attributes of reducing periodic fishing pressures on fish stocks, it is very difficult to enforce without sophisticated communications and data networks backed up by an expensive, real time, accurate and verifiable monitoring and surveillance system. This management scheme may be appropriate for developed countries, but could prove very difficult to implement in developing countries. Experience has shown that the ITQ management scheme may be appropriate and manageable in small island states with both small populations and fishing fleets, and where the fishery is largely export oriented. To date the system has not been found appropriate in other situations, especially where the fishery has a domestic component.

An initiative which will result in a near, real time vessel monitoring system is being developed by the FFA for use in the South Pacific. Until this system is perfected and acceptable to the fishers, it is suggested that such a quota management system may be found to be administratively difficult to operate and result in a continuation of overfishing.

It has been found over time that no MCS activities will be successful if there is not an understanding and acceptance by fishers of the rationale behind the MCS actions being implemented. Legislation which is unenforceable denigrates the credibility of the entire fisheries management process in the eyes of the supposed benefactors, the fishers, and consequently quickly establishes a "them and us" confrontational attitude on both sides. It must be remembered that most fishers are realists who are fishing for their survival and "whilst sometimes libellously assumed by the ill-informed to be crooks, are perhaps best described as being as honest as the next man, but hard, individualistic businessmen running very competitive and often highly capitalized operations. It is worth remembering that they do so in the face of a largely unforgiving sea that creates a working environment which, in terms of industrial health and safety, [has one of the worst industrial accident rates in the world]: and they operate increasingly in an economic climate of ever increasing overheads countered only by the proceeds of catches which [are] subjected to [greater] quota restriction. All of this is done in the knowledge that the success of their venture and the livelihood of their crews depends entirely on their individual skill, effort and initiative. Given these pressures, it is perhaps not surprising that such independent minds do not always take kindly to bureaucratic controls, especially if these appear to them to have little practical purpose"

Components of MCS

Fisheries monitoring, control and surveillance (MCS) is a key feature of the fisheries management process for which FAO organised an expert consultation in April 1981. Although the objectives of fisheries management and MCS are generally to take advantage of the economic opportunities of the extension of the Exclusive Economic Zone (EEZ), they also include the exercise of sovereign rights over the zone, conservation of marine resources, and collection of appropriate data on activities to ensure sound, rational oceans and fisheries management planning.

There are three main components to MCS which, depending on cost, commitment, and organisational structure, will be configured uniquely for each system. These are the land, sea and air components. The latter now often includes the use of satellite technology. The land component, or base of operations, can serve the inland, freshwater, and coastal aspects of fisheries monitoring, control and surveillance. The land component is usually the coordinating sector of all MCS activities and regulates the deployment of available resources to best address the changing situations in the fisheries. The coastal/land component is the sector responsible for port inspections and the monitoring of transhipments and trade in fish products to ensure compliance with fisheries legislation.

The sea component of MCS includes the actual technology for surveillance of the national, sub-regional or regional maritime zones of control. This component can include radar and vessel platforms which are utilised for these purposes. Traditional apprehension of an alleged violator of the laws which apply in an EEZ requires a "laying of hands" on the offender, mainly for the legal formality of arresting, but also for identification and securing of evidence. As this is sometimes costly in terms of vessels, crews and supplies, many nations are now favouring "no force" surveillance techniques.

These include the use of observers, national or regional vessel registers or licenses and agreements which include clauses regarding the responsibility of the flag State for the actions of its citizens and vessels. The fisheries management strategy, if it utilises zonal fishing divisions, quotas necessitating catch monitoring, mesh and gear restrictions and minimum/ maximum fish sizes, requires vessels to carry fisheries personnel which can remain at sea with the fishing fleets to perform these functions, as they cannot be expedited from land. The patrol vessels also serve a maritime safety function while at sea.

The air component of MCS is usually the first level of response to a coastal state/regional concern in its area of responsibility or interest. The flexibility, speed and deterrence of air surveillance makes it a very useful and cost-effective tool for fisheries management. This component also provides the cheapest and most rapid information collection on fishing effort in the zone of interest, from either the aircraft or satellite platforms. The cost of these systems is directly related to the sophistication of the technology utilised. Air surveillance, while providing initial information regarding the activity in the fisheries, can also be the first indicator of potential illegal activity in the zone. This latter information is the base on which further MCS action can be precipitated. Air surveillance also has the added advantage of a secondary tasking capability for fisheries habitat and general coastal zone management monitoring.

The potential benefits to seafarers in difficulty at sea, pollution, habitat and general coastal zone management monitoring are significant in that they can also be addressed during fisheries MCS activities. Multi-tasking of expensive fisheries MCS resources for other fisheries-related monitoring functions can be cost efficient and also effective in terms of integrated ocean management programmes, in particular with respect to fisheries and the marine environment. Acceptance of the aforementioned definition of monitoring, control and surveillance emphasises the point that MCS is a three-tiered system. A common error in establishing MCS systems throughout the world is the concentration on the "S" in MCS, or the enforcement phase, without due regard to the importance of the other two phases.

The monitoring and control aspects of MCS provide the base of information and legal framework for sound fisheries management and operational planning. The surveillance phase is the most expensive aspect of MCS and hence, developing countries must look at the most cost-effective methods to carry out functions related to this component. It is useful to remember that the most beneficial aspect of enforcement is preventive enforcement, or voluntary compliance. It is similar to car repairs where it is less expensive to conduct preventive maintenance than to carry out time consuming, and costly, repairs. The same principles apply to MCS.

An understanding and acceptance by those who will most benefit from MCS actions, the fishers, can prove most effective in gaining their support, commitment and involvement for surveillance purposes. Voluntary compliance has the added advantage of then permitting a focus for the expensive enforcement resources towards areas of major national, sub-regional or regional concern in the most cost-effective manner. The deterrent impact contained in the control phase, or the legislation supporting MCS, will determine the potential repetition of offenses. If there is an understanding of the rationale for the fisheries legislation, this will disuade the fishers from violating these acts and regulations. This idea can be reinforced through the drafting of enforceable legislation and appropriate penalties.

There are three linked components of fisheries management:

1. collection of data on biological, economic, social aspects of the fisheries and basic information on the fishers, boats and gear;
2. decision making or fisheries management planning; and,
3. implementation, the MCS aspect of fisheries management involving both government officials and members of the fisheries community and industry.

The first component includes collection and analysis of biological/resource assessment data, basic fisheries information on fishers, boats and gear, fishing trends and patterns, and social and economic data for the harvesting, processing and marketing sectors of fisheries. The analysis of these data provides the input into the fisheries planning exercise.

The second component, includes the entire process of consultation and negotiation with all parties which influence the decisions concerning a fishery. This should result in fisheries management plans for the harvesting, processing and marketing sectors. The key factor in this decision-making component which affects the entire process of fisheries management is the political will and its commitment to sustainable and rational fisheries management. Political commitment is crucial to success in the implementation of fisheries management plans.

The last component, which is often the most difficult for governments to deal with due to potentially high cost or other government priorities or arrangements in the fisheries sector, is the implementation of these plans. It is represented by monitoring, control and surveillance of the fisheries and the fishers, and is an absolute requisite to the successful implementation of fisheries management plans. The most comprehensive and acceptable plan on paper will not result in successful fisheries management unless it is implemented through the use of monitoring, control and surveillance operations. Lack of attention or commitment to implementation of MCS activities often results in overfishing, collapse of resources and economic loss to future generations.

In the past, officials often reflected the view that fisheries management includes only the biological studies for resource assessment and development of management plans, and there the process ends. The fish were then expected to conserve themselves and the industry was to be the steward for processing and marketing. The support for data collection and policing of fishers and the industry to ensure appropriate input into future fisheries management plans, and the successful realisation of these plans, was low. Now, there is a growing awareness of the declining condition of the environment and a greater acceptance of the need for investment in implementation (MCS) of natural resource management plans, including those for fisheries. It is in this vein that countries are attempting to address the earlier concerns with respect to the funding of MCS activities.

As fisheries have the greatest risk with respect to mismanagement of renewable marine resources and their habitat, it may be a consideration that fisheries departments be delegated this lead role in MCS matters. As noted earlier, however, MCS activities in support of fisheries management can also accommodate other secondary tasks which pertain to conservation of a nation's natural resources and its environment. The expensive infrastructure required for MCS activities, especially in terms of aircraft and vessels, when coupled with the overall responsibility for MCS of the coastal zone and EEZ of each country, or group of countries, for fisheries and their habitat/marine environment is easier to rationalise as being appropriate to the magnitude of the task.

Design Considerations

Influential Factors

There are three groups of factors which may influence decisions respecting the type of MCS system required to meet the needs for fisheries management. These include the geographic, demographic, industry and international profile of the fishery, economic factors, and the political will and commitment related to analysis of the first two groups of factors.

Size of the EEZ and the Fishing Area within the Zone: The area of fishing of both the domestic and foreign fleets will have a significant influence on the design of the MCS system for each country. There will be a significant difference in the situation, and hence the MCS strategy, in the Philippines, where responsibility and authority for the artisanal fisheries has been delegated to the municipalities, compared to that for a country such as Equatorial Guinea, with 8,750 artisanal fishers and foreign fishers taking advantage of the State's inability to fund and operate appropriate MCS mechanisms.

The area of active fisheries may also raise control concerns, due to the migration of fish stocks between fishing areas and countries. This can also cause an artisanal versus offshore conflict, if there are incursions of the latter into the small boat fisher's area. This can become critical when the EEZ of a country has only a small rich fishing ground and the fishing pressures from all fishing sectors is intense. Another consideration in this regard is the size of the various fisheries. If the domestic fishery is small, inshore stocks are strong, and the fishery is not a threat to conservation, then MCS activities can focus on data collection and reasonable control mechanisms required to maintain the health of the fishery. If the offshore fishery is extensive and lucrative, involving either domestic or foreign-owned vessels, then MCS activities to collect data, control the fishery, and patrol the area may require greater effort, to ensure that benefits of the resources are conserved for the State.

On the other hand, if the artisanal fishery is overfished and there is growing pressure on the offshore fisheries, with little knowledge of the resource base, then consideration will need to be directed to greater information gathering on the latter and options for re-directing effort from the artisanal to the offshore fisheries in a cost-effective, controlled manner, if this is possible. A key factor in this scenario may be the targeted species, both offshore and inshore, offshore usually being high value species for profit while the artisanal fisheries focus on any fish, often lower value, for survival.

The strategy to establish an MCS system for a foreign fishing zone that is restricted outside 12 or 15 nautical miles is much easier to develop than one in which foreign fishermen are permitted in inshore zones when fishing for certain species. The latter necessitates a verification of fishing catches while the former only requires geographic confirmation. This is one example of the kind of strategic analysis which is required when developing an MCS plan appropriate for the fishery and the enforcement capability of the State.

The physical size of the EEZ and the active fisheries area within the zone will have a significant influence on the design of the MCS system. A large zone and fishing area may require air surveillance or other such infrastructure to patrol the areas of concern, whereas a narrow fishing zone might be surveyed cost effectively using other technology, possibly land-based, such as over-the-horizon radar, coast watch systems, or vessel monitoring systems (VMS), coupled with less expensive "no force" strategies. An example of this difference is the comparison of the 209,000 square kilometres of EEZ for Ghana, compared to the 20 million square kilometres of EEZ to be surveyed by the FFA for its member countries. The options for MCS strategies must be considerably different if the systems are to be cost effective and this depends largely on the size and geographical location of the fishing areas within these EEZs, or collective areas of responsibility.

The profile of the fishing fleets, domestic, foreign, artisanal and offshore, is a consideration for the implementation of MCS strategies. The age, condition, size, fishing capacity, gear type and fishing patterns of the vessels will all have an impact on what the State may wish the vessels to do to comply with its MCS policies. These factors may also pre-empt requirements to carry certain equipment, due to the state of the vessel, crewing, current equipment or other such factors which can increase costs to the point that fishing may not be viable, especially for the domestic fleet.

On the other hand, there may be a requirement, after noting the profile of the fleets, to establish minimum safety and equipment standards not only for the well being of the fishers, but also to minimize the risk of pollution at sea.

Topography of the coastline: A coastline with several bays, river outlets and important mangrove habitat would require a more complicated MCS scheme to conserve fisheries resources than one with steep rocky cliffs and minimally important habitat. The difficulties for surveillance would also be more complex in the former case, due to the indentations in the coastline which would necessitate a physical presence to survey fishing activity, rather than radar or other, less expensive technology.

Other interests: The importance of tourism, the enhancement of industrial capacity, the requirements for ports and shipping, maritime safety-at-sea and pollution monitoring and controls can all have an impact on the strategy developed for fisheries management, with consequent repercussions on the MCS system adopted. There will be a requirement for discussion and liaison with appropriate ministries to ensure that government priorities are met and, in so far as possible, fisheries requirements and benefits to society are recognised and respected. Fisheries management priorities will often seem to conflict with priorities from tourism, industry, and marine transportation initiatives. It must be realised however, that the sustainable benefits of each of these industries, and the economic and employment situation for the State, will benefit from appropriate attention to, and conservation of, marine resources. There will be a need for a mechanism for discussion and resolution of differences in approaches and priorities for each of these important industries.

International pressures: International pressures from distant from water fishing nations (DWFNs) and the short term economic benefits of foreign currency cash flow can be attractive to states that do not have fully developed fisheries, but often with the consequence of non-sustainable exploitation of their fisheries resources. DWFNs have their own difficult situations of over-capacity and recession in their economies and fishing industries which they have been trying to address. These countries may, in the past, have been be less appreciative of, or sensitive to, the negative impact of some of their agreements with the developing nations. On the other hand, it would have been unreasonable to expect that these fleets would negotiate an agreement which would have had them operating at a financial loss. Consequently, these agreements have in the past, unfortunately, been to the detriment of the conservation goals of coastal states, which did not, and still do not, have the economic flexibility to ensure the implementation of appropriate MCS systems to conserve their fisheries resources.

Often, the only protection against uncontrolled overfishing and lack of compliance with regulatory measures due to shortages in resources and training, is through strong co-operation on a bilateral, sub-regional and regional basis with regard to MCS activities. Cooperative efforts can result in economic and international pressures against noncompliance with internationally respected, conservation principles which would not otherwise be achievable on a single state basis. The potential for the establishment of international standards and guidelines is greater in this forum of international and regional cooperation. Examples include the aforementioned initiatives such as; the flagging agreement, Code of Conduct for Responsible Fishing, port state control and the control of fishing for highly migratory species and straddling stocks.

In recognition of the rights and interests of all States under the Convention on the Law of the Sea, developing countries must address their obligations to deal with conservation concerns and, at the same time, international fishing partners must ensure compliance with national fisheries legislation and international conventions dealing with fisheries in the name of conservation. One particular note of special pertinence today, is the economic temptation to register DWFN vessels in national registers when there is no capability to control the activities of the "new flag" vessels. Some of these "new flag" vessels operate with short term immediate interests in internationally sensitive areas of the world and without appropriate attention of conservation. This brings international pressure on the flag State with respect to its credibility and commitment to internationally accepted fisheries conservation and principles and to the implementation of the Convention. Registration of these vessels should be avoided.

Involvement of fishers, communities, organisations, cooperatives, unions and fishing companies: It should be self-evident that the cooperation of the fishing industry and fishers is essential to cost effective fisheries management. If the industry, fishers and their communities and organisations actively participate, and are recognised as real participants, in fisheries management, MCS planning and related activities, there is a much greater potential for successful implementation of these plans, to the benefit of all.

On the contrary, if the opposite transpires, it will be extremely expensive and very difficult to successfully implement any fisheries management plans. Lack of recognition, input, involvement and understanding of the principles and rationale behind the proclaimed fisheries management scheme has often resulted in non-compliance, alienation of the fisheries department officials and active subversion of the intended plan, thereby placing much more pressure on the need for surveillance. Most fishers wish to conserve the resources and will support such initiatives if the efforts are reasonable, enforceable and understandable.

MCS operations directed towards education and seeking input from fishers are facilitated if there are strong fishers' or community organisations in place to discuss these issues. The independence of fishers is well known, consequently there is often a reluctance for fishers to join together for these types of discussions. Assistance to fishers in getting them to recognise and accept the advantages of having a collective voice is one of the challenges of fisheries educational initiatives in seeking input and support for MCS activities.

Demographics of the domestic fishery: Other issues may impact the type of MCS strategy to be adopted to ensure the continued health of fisheries, especially the domestic fishery. One such example is in the Seychelles, where the demography of the fishery is such that the average age of fishers is very high, as other employment opportunities appear more appealing and lucrative to the younger population. This creates several unique challenges for the State to re-kindle an interest in fishing as a profession and to build a controlled fishery using Seychellois, instead of other international fishers. This also presents an opportunity to educate the new fishers in the benefits of fisheries conservation. Alternatively, if young Seychellois are not encouraged into the fishery, the MCS system design will need to focus on the possibility of an increase in foreign offshore fisheries.

Economic Factors

Contribution to the GNP: It is obvious that the contribution of the fishery to the national economy will determine its profile and the importance placed on fisheries management activities. It makes economic sense, however, that the benefits from the resources exceed the cost of their conservation. As seen in the Seychelles, the potential contribution to the national economy by the displacement of international fisheries with a corresponding increase in domestic fishers, or other offshore fishing arrangements, could enhance the current situation. The GNP factor will likely have a significant effect on both the scale and the design of an MCS strategy.

Earnings of foreign currency: A factor of considerable importance to several developing states is the earning of foreign currency by permitting international access to the fishery. It is very unfortunate that certain distant water fishing fleets have exploited this need in a strict business sense, to meet their own financial and employment requirements, without due consideration for conservation, or reasonable returns, for the coastal States. In the case of a developing country, where trained resources and infrastructure are not available to ensure the implementation of adequate MCS strategies, this can become a serious problem in the conservation of their fisheries resources. This is sometimes further complicated with the offer of external financial incentives to subvert any MCS efforts that the State may wish to impose. Governments, both developing and developed, must ensure that the fishing opportunities granted for both domestic and foreign fleets result in appropriate levels of compensation, and that these funds benefit the State.

Employment opportunities

The employment opportunities which can result from enhancing the fishing potential of the coastal State are also a factor in the consideration of the MCS strategy. In a situation where there are seen to be advantages in the long term of displacement of international fishing fleets, this may require training of coastal or island State nationals who will eventually assume these fishing rights. Training of nationals could therefore be a component of the access agreement with third party fishing fleets.

The demonstration of an increase in benefits to the individual fishers, either as increases in income, or employment opportunities, will have a positive impact on the national economy, the strength of the fishing sector and support for government fisheries policies. Linked with the above strategy of training of nationals could be the opportunity to ensure the implementation of appropriate safety-at-sea equipment and practices in accordance with the coming into force of the Protocol to the Torremolinos Convention 1977. This Protocol will bring fishing vessels under port State control with respect to safety certificates and early attention to this point can result in economies for future training. This initiative can also link to the development of a Code of Conduct for Responsible Fishing resulting in a new attitude and blend of fishers with conservation and safety high on their list of priorities.

Conservation of Fisheries Resources

Recognising that MCS should conserve fisheries resources and their habitats, there may be mutual benefits for other ocean users as well as the fishers, if appropriate liaison between these users and the fisheries department and relevant MCS strategies can be developed. For example, careful assessment and control of tourism development, assurance of non-destructive fishing practices, development of marine parks, use of mooring buoys to reduce the damage to coral reefs from ad hoc anchoring, etc. can all benefit both fishers and other ocean users. MCS strategies can be developed to include consideration for such activities.

Small island States are realising the negative impact of excessive use of pesticides and the marine pollution resulting from uncontrolled industrial development. It is being noted that all land based activities on small islands eventually influence the marine environment and they have the potential to kill the very marine resources and habitat, including the coral reefs, which bring the tourists and foreign currency to the State. Belize, in Central America, has been attempting to regulate tourism, the development of marine parks, and the fisheries, and has established appropriate surveillance initiatives to ensure the implementation of the required management plans.

Other secondary benefits from MCS activities outside of fisheries activities can also be factors in establishing the type and structure of an MCS system. MCS activities can also assist in addressing safety-at-sea for national and international seafarers and marine pollution, environmental matters.

Low cost protein: A further economic factor to be considered in the design of an MCS system is the requirement of the State for protein for its citizens. The MCS design can include the requirement for a percentage discharge of fisheries products in the coastal State for distribution, processing or further export. This could result in the direct provision of protein for the people, or enhance industry development in the fish processing sector and have resultant positive impacts on export earnings for the procurement of this protein.

Political Will and Commitment

The key behind any ocean policy, planning and management system, including that for fisheries, is the degree of political will and commitment to implement such a system. The following includes some of the factors which may form the base for decisions on the structure of the MCS system to be developed. The economic profile, or potential thereof, of the fishery in the national economy will undoubtedly determine the level of support the MCS initiatives will receive from the government. A potentially lucrative domestic fishery, and the MCS activities required to protect it, will probably receive significant government attention.

Nevertheless, it will be necessary to balance the long term benefits with the short and medium term benefits to maintain the political support which is key to the successful development and implementation of MCS systems. Some of these could include the establishment of a database for resource management, revenue from resource users through license fees and greater control of the resources from licensing and surveillance, which conserve fish stocks and hence, increase incomes of the fishers. It must be emphasised that the political will and commitment of the country is the key factor in the successful design and implementation of MCS systems for fisheries.

This set of influential factors and their relative importance to the political objectives of a country make fisheries management and the resultant MCS strategy unique to each country. Key observations from reviews of successful national and regional MCS include:

1. There are no models and each system [in operation] is in fact, adapted to the cultural, geographic, political and legal framework of the country or region.
2. The operational character of the system will depend on management decisions made.
3. The legal and policy considerations are always taken into account when establishing an MCS system.
4. The decision making power is always in the hands of the civilians, even on surveillance matters.
5. The national and regional MCS are complementary to each other.

A growing trend over the past several years has been the formation of groups of countries with common political and economic interests. An enhanced awareness of the environment and the requirement for the conservation of natural resources has brought recent activity and interest in the formation of associations and international agreements along regional lines. The rich resource base in many developing countries, their less than favourable economic situation and inability to protect these resources and the desire of the international fishing industry to gain access to these resources, all emphasise the need for cooperation amongst developing fishing nations. There are proven cost savings which can be experienced through cooperation with respect to acquisition of MCS resources, training, and negotiating from a larger power bloc for reasonable compensation in return for access to underutilised resources. The international exchange of appropriate fisheries data for MCS and fisheries management purposes, harmonised legislation, extradition and port State agreements are all benefits which should encourage international affiliations within regional developing countries.

The decision with respect to establishing an MCS system on other than a national basis does, however, depend on several factors. These include whether there is an existing organisation which will serve the purpose, whether there is the international political will of states in the area, common interests in fisheries which would benefit from such a liaison, common language and cultural ties, and whether differences can be overcome. The security of what may be considered as sensitive data and the potential to resolve internal concerns to present a common face to the outside world are all additional considerations which impact on this decision regarding international fisheries and MCS cooperation.

Finally, the difference in economic situations of possible member countries and the cost sharing arrangements for support to an international organisation are also critical factors to be addressed. Each country must balance their own advantages and disadvantages prior to making a commitment to regional cooperation. Probably the most advanced of such organisations in the developing world is the South Pacific Forum Fisheries Agency, with sixteen members, in its second decade of cooperative fisheries management. Another example is the Organisation of Eastern Caribbean States Fisheries Unit, a sub-regional organisation in the Caribbean Basin.

A sub-regional initiative is commencing in the Indian Ocean through the Seychelles which will link with the regional tuna research programme in Sri Lanka. Efforts of the CARICOM Fisheries Resource Assessment and Management Programme is another example with considerable potential for success under CARICOM management. Efforts in East Asia with fisheries management and MCS are well advanced and recently, operational manuals have been developed. These organisations can provide additional information for the consideration future MCS strategies.

A decision with respect to international cooperation will not abrogate the State from its responsibility to establish appropriate structures internally to address fisheries MCS issues. As an MCS system is developed, it is expected that there will be interest from ministries other than fisheries for access and input into the priorities and tasking of the resources. The ministries responsible for environment, national defence and coast guard, customs and immigration are a few which can be expected. It has been noted that MCS surveillance resources are expensive and that multitasking could be cost effective and efficient. Experience has noted however, that too many priorities can result in the acquisition of capital equipment which does not meet any function appropriately, consequently, it is suggested that for fisheries MCS activities, coordination be with other ministries with fisheries-related interests, such as coastal zone management and the marine environment.

There is also a very real requirement to recognise that the ministry, or department, with a considerable stake and interest in conservation and sustainable use of ocean resources and their habitat, is fisheries. There is also a need in any operation to ensure that there is one lead agency with the appropriate authority to make decisions. Although there are several ministries with interest in MCS and, consequently, there may be a need for a coordinating committee, there still needs to be one final authority for decisions on the deployment and priorities for MCS operations. Split operational "command and control", to use military phraseology, have not met with success in the past in military or civilian operations. It is recommended that the fisheries department be provided with the lead role and ultimate authority for ocean sector MCS activities, in consultation with other interested departments and ministries. If this is not acceptable to governments, an alternative could be an alternating chair for the coordinating committee, but this is a less preferred option. It is always preferable for fisheries managers to have to report to only one superior to maximise efficiency in operational MCS activities.

Infrastructure Requirements

The control component of MCS will necessitate the determination of appropriate and enforceable legislation required to implement the fisheries plans for the various fisheries. It will address the authorities of fisheries personnel, legal fishing activities, minimum terms and conditions for fishing, and penalties for non-compliance. The minimum conditions which a State may wish to implement could include vessel identification, catch and reporting requirements, conditions for transhipment, standard catch and effort log sheets, terms and living conditions for observers, local agents for international fishing partners, and flag State responsibility for their vessels.

The control component will link with the State's justice department and also necessitate the appropriate training of all personnel involved in enforcing the legislation, including sessions where the assistance and advice of the judiciary is requested. The infrastructure requirement for this component of fisheries MCS is a team of knowledgeable fisheries lawyers for both drafting of enforceable and appropriate legislation, and also for the legal follow-up which may be required to implement these laws. This component will also establish the various mechanisms, strategies and policies for the implementation of operations (MCS activities) to implement the fisheries management plans.

The surveillance component of MCS will require fisheries personnel who not only collect data for the monitoring aspect of MCS during their surveillance duties, but also have the appropriate equipment, operating funds and training to enforce the legislative mechanisms of fisheries management. These personnel will require direction and infrastructure from which to operate, be it land, sea or air facilities. This is the enforcement component of fisheries MCS and as such is usually the largest and most expensive activity to fund. It must be remembered that for international MCS activities, there is a requirement under the Convention on the Law of the Sea that all surveillance equipment be clearly marked and identifiable as on government service.

A central headquarters near the departmental decision makers for the coordination of fisheries operations. This headquarters, as well as having the offices for the administration of fisheries, would, ideally, be situated adjacent the operations room. A central operations room where the current status of the fishing operations can be shown through maps, plots and computer enhancement is recommended. This centre would need offices and personnel, with communications to appropriate field offices and other enforcement agencies, and direct communications to the minister responsible for fisheries. This becomes the situation briefing and de-briefing room when a sensitive fisheries matter arises.

The Fisheries Administrator should thus have the capability, through the equipment and information accessed from this centre, to show the situation to decision makers and obtain direction for timely responsive action. These centres can be staffed by as few as two persons trained in communications, computer access and display techniques. The communications system would ideally have telephone and appropriate radio communications to all fisheries centres and mobile platforms in the field for both safety and control of operations. Some MCS systems also use satellite communications in their networks, but it is very expensive. The modern HF radio systems on the market today could possibly assist in keeping costs to a minimum without losing effectiveness.

The computer data system for licensing and vessel registration, if it is decided to use such a system for this basic information, is now affordable. There are several licensing and vessel registration systems in use today and it will be a decision of the Fisheries Administrator as to the system which will best meet the State's needs. It is anticipated that the procurement of other major MCS equipment will be coordinated from the central headquarters. The air surveillance requirements for MCS may appear expensive, but are still seen as the cheapest method to receive rapid surveillance information with respect to fishing and fish habitat information in the zone.

As a minimum, the following equipment is recommended. It is highly desirable to have a twin engine turbo prop aircraft for over-sea flights for safety, endurance and low maintenance costs. As a minimum, this aircraft should have radio communication to base and directly to sea-going fisheries patrol vessels, with common marine frequencies. The navigation system of the aircraft will need to be accurate, for it will form the base for prosecution of area infractions if they proceed to court. It would be desirable to have an endurance capability of 4-6 hours at economical speed. The speed for transit should be reasonable to maximise the time in the assigned patrol area, but the aircraft should be able to go slowly enough at low levels to identify and photograph fishing vessels. Photographic equipment for the recording of vessel activities is necessary.

Night lights and instruments for IFR flights are very desirable for surveillance purposes. Onboard computers linked to accurate navigation systems, communications systems, radar and photographic systems, and the capability to accommodate vessel monitoring systems, would result in a very technologically advanced air surveillance platform. This would however, be an expensive operational tool which might be inappropriate for budgets of developing countries. The expense could possibly be easier to accommodate through regional cooperation and shared use of the air surveillance equipment.

The choice and equipping of the aircraft will be dependent upon the cost of the airframe and the ongoing costs of operation and maintenance. These latter two factors are often lost in the considerations for air surveillance, but they are the most significant costs for the MCS air activities. Aircraft can cost from a few hundred to several thousands of dollars per air hour depending on the configuration of the craft and equipment. If at all possible, it is highly desirable to ensure that there is local access to appropriate training and equipment to permit long term maintenance of the aircraft.

The sea-going requirements will vary considerably between countries, depending on the MCS strategy. The primary consideration when considering the acquisition or use of patrol vessels should be cost-effectiveness and affordability for the primary task, fisheries surveillance. One golden rule for cost effectiveness is that the fisheries patrol vessels should have at least the same sea keeping capability as the fleet which they are monitoring. There may be a temptation to procure fast, expensive vessels; however, it must be remembered that the purpose of these vessels is to transport the authorised fisheries officials to the fishing vessel for inspection. Although desirable for a quick transit to the patrol zone, this capability must be balanced against the high fuel and maintenance costs for such machinery.

There may be a requirement to be able to overtake a departing vessel where there are no other diplomatic arrangements in place, but this capability should not overshadow the need for staying at sea and cost effectiveness on a daily basis. Fast patrol vessels have a tendency, no matter how well trained the operator, to operate at near top speed making them expensive in fuel and maintenance, with long down times due to equipment wear and repairs. Economic considerations may also make vessel charters a viable option instead of purchases. In this manner maintenance becomes the concern of the contracting firm and not the government.

Coastal and nearshore vessels do not need to stay at sea for prolonged periods and, consequently, smaller patrol vessels with one or two days sea keeping capability, or rapid response shore-based craft, might serve the purpose in this latter case. These vessels would be best equipped with radar and communications systems. The latter should include both marine radios and an additional one to communicate with the air surveillance platforms. Equipment for boarding and an appropriate boarding craft are recommended. Most countries have found the fibreglass, V-hulled, rubber-sided, speed boat to be most effective for boarding. The boarding boat should have two outboard engines, or one inboard/outboard and a small outboard engine, for safety. The boarding boat requires communication equipment to remain in contact with the patrol vessel at all times.

The offshore fishery will require the largest, and hence the most expensive, sea-going platforms in the infrastructure for fisheries MCS. These vessels can range from deep hulled trawler type vessels to offshore oil supply vessels with helicopter landing facilities. The key in the choice is, again, the capital cost for the vessel and equipment and, equally important, the operating and maintenance costs. Large vessels, by their very nature, require considerable fuel and provisions to operate for extended periods at sea. It has been recommended that wherever possible the management strategy attempt to keep the need for these expensive seagoing platforms to a minimum, but it must be realised that they are necessary for most traditional fisheries management schemes.

Most offshore vessels for fisheries would best be equipped with twin diesels of a dependable model, with trained engineers, up-to-date navigation equipment, radar, photography equipment and radio communications. The latter should be at least as per the inshore patrol vessel, preferably with back-up systems and ideally, linkages to the air surveillance platforms. These vessels are intended as boarding platforms and their regular duties should not require them to be heavily armed assault vessels. As noted earlier, their first role is as boarding platforms for the fisheries officer.

Office space is required for the field staff and their supporting administration. The office should be equipped with communications equipment to maintain contact with the headquarters and also to maintain communication with staff while on patrol. A radio communication network is usually sufficient for these activities. The office also requires the capability to collect and transmit data to other offices for compilation and analysis and also, to receive results for the planning of operations. Ideally, this capability can be achieved through a computer system with communication to these other offices. Transportation is required for staff for patrol purposes, either along the coast, at sea or by land, and also along the rivers and lakes where there are active fishing operations. This transportation can range from small boarding type craft, to motorcycles, to other types of vehicles. It is highly recommended that staff patrol in pairs for safety and personal security.

It is assumed that the Fisheries Administrator will ensure that each field office has in its reference library certain documents for assistance in their duties. These include:

- — current fisheries legislation, acts, regulations, notices and the gazette,
- — departmental guidelines for MCS activities including those for prosecutions,
- — copies of any applicable treaties or agreements between countries in the region,
- — a set of charts with updated baselines, territorial seas, EEZ and any specifically noted areas for fisheries management,
- — past fisheries cases, details and penalties for reference during the preparation of a case.
- — safety procedures and guidelines for MCS.

Each officer should have in their possession, at all times, pictured documentation which clearly identifies the individual as a government authorised fisheries officer. This requirement should also be in the fisheries legislation. Each officer requires communications equipment to maintain contact with their base of operations. Each officer must have the appropriate accoutrements to record findings during the patrol; e.g., a patrol book with clear identification of the owner and sequentially numbered pages. This notebook could be used in court proceedings for identification of events and as an aide memoir for the officer. It is essential that it is properly maintained.

A final item for careful consideration is the provision of firearms to staff. There are several considerations with respect to this matter but in general firearms, if they can be avoided, are not recommended for fisheries MCS. However, it is recognised that there are situations when it would be considered very dangerous for fisheries officials to conduct their business without adequate personal protection. These include the proclivity of firearms use in the country, in the fishing industry, and levels of illegal activity throughout the industry and at sea. It is also important to assess compliance trends in the fishing industry and the history of difficulties with fishers, both domestic and foreign, regarding the protection of fisheries staff. Where fisheries have become incontrolled and unmanageable, it has been found that fishers resort to other, less desireable and violent activities, thus making the protection of fisheries MCS staff a priority requirement. Where possible however, other means of protection, such as guard dogs or batons for land based fisheries personnel, are urged. The issuance, carriage and use of firearms should be considered or as a tool for staff protection only; it is not recommended that firearms be considered or used as an aggressive surveillance tactic.

If it is therefore decided that firearms will be issued to staff, new considerations apply. The first of these is the appropriateness of the designated individuals for the carriage of the weapons. Not all persons are mentally suitable to carry and use firearms. The Canadian experience in fisheries recognised this fact and now a battery of psychological tests, as part of the selection process, are used to screen applicants for their suitability to carry firearms. Those found not suitable are released from further recruitment testing. The danger in putting weapons in the wrong hands in terms of potential accidents can be significant and result in legal liability of the Department. The second major consideration is the initial training and the need for ongoing refresher training. This is critical for the confidence of the staff in the proper use of the weapon and their own personal safety.

The decision on arming vessels for fisheries enforcement purposes is one which should not be taken lightly. It is a conscious decision to arm the vessel both for protection, and potential aggressive action. This decision may be necessary where fishing vessels commonly do not comply with the orders to halt for fisheries inspections. This scenario may result if no other enforcement strategies or agreements to ensure compliance have been established with the flag State of the vessel and diplomatic relations to address the situation are not available or have failed. In this case it may be the government's policy to permit aggressive, civil police action to apprehend the alleged offender.

The matter of consequence here is the use of aggressive force. Legally, this force should be limited to that necessary to ensure compliance with the legal authorities, in this case the fisheries officer, or for protection of staff as appropriate. The amount of force that is deemed appropriate is always one of subjectivity, but excessive force may result in a case, if it goes to court, not being supported. It is suggested, therefore, that in the case of action involving the use of fire-power, there be strict standards of application for the escalation of the use of force, from verbal warnings, through earning shots, to the use of force to physically stop the vessel; i.e., stop or potentially, sink the vessel. This action also includes the assumption that the boarding vessel is appropriately identified in accordance with the obligations under the Convention and that verbal identification has also been passed to the vessel being boarded.

Countries should ensure that appropriate higher authorities are involved in the decisions to escalate the use of force. The use of force for protection when fired upon always remains with the master of the vessel to protect the ship and crew. The use of firearms by a fisher against a fisheries official or vessel, noting the identification above, should be responded in kind, with the maximum use of force directed at the source of firing. It must be remembered that MCS surveillance is essentially a civilian police action, and not an aggressive military exercise hence, the use of military forces for fisheries MCS is recommended only for extreme cases.

This section has been presented to encourage reflection on the initial decisions regarding equipment to implement the MCS operational plan. The sum total is a requirement for a central control facility, a data system and network to the coastal areas and offices, offices along the coast for data collection and surveillance, communications and data collection facilities, appropriate transport equipment, land, sea and air platforms, and other safety equipment for surveillance operations. The requirement for appropriately trained directors, supervisors, and field personnel is assumed to be a central component of MCS.

The requirements for MCS include personnel to address each of the components of monitoring, control and surveillance. The numbers of these personnel will vary with the MCS scheme in place, but basic requirements for the qualifications of these persons should remain fairly constant. These personnel need varied levels of expertise; for example, data collectors need to be literate, have good interpersonal skills and knowledge of the fishery and its policies and procedures. Sea-going experience is crucial to fisheries MCS, especially for sea operations. These individuals are often at the technician level. Observers for offshore vessels also fall into this category of staff. It must be noted here that the observer scheme is only appropriate if capable, honest and dedicated personnel are available, preferably with offshore sea experience.

In many countries, close supervision is also necessary for this scheme to be productive due to the onboard pressures placed on these individuals. If these factors cannot be met, consideration for the implementation of an observer programme is inappropriate. The ability to analyse the data collected for fisheries management decisions and operational deployments requires a higher level of knowledge and competence, both academic and practical. The control component requires individuals with comprehensive and very good knowledge of fisheries and the law. These individuals would work with the lawyer assigned from the Ministry of Justice for the design of enforceable laws and also for the internal decisions regarding MCS operations.

Finally, there is the need for surveillance personnel. This includes personnel to operate small and large patrol vessels. Further, aircraft personnel will be required, as will the every day fisheries officials at the junior and more senior levels for coastal, river and lake patrols, management of the offices and liaison with the fishers. The sea-going surveillance personnel should be recruited from fishing communities, where appropriate personnel can be found. The air surveillance personnel can possibly be seconded from the military, or possibly local airlines to minimise training requirements. Support personnel will also be required from local staffing pools. Maintenance personnel for equipment will need to have experience, or be trained, on the equipment provided. The training for fisheries technicians and officers is key to ensuring competent staff for the implementation of fisheries plans.

Impact of New Technology

There can be a tendency when considering a new system to look for the most advanced technology which can do the work required. This is usually also the most expensive equipment on the market. Keeping in mind the economic logic that the overall expense of a conservation system should not exceed the benefits gained from the fishery, then it is more prudent, especially in the current economic climate, to look for appropriate and affordable technology for each fishery situation.

The requirements for data collection, and verification, usually include information on the fishers, their fishing vessels and gear, their individual and community dependence and returns from the fishery and hence, the period, quantity and value of their landings. Further information on the area of capture and the size, weight and age of the fish are also beneficial for fish stock assessment modelling and stock predictions, despite their inaccuracies. It could follow then that the most cost effective strategy would be to have data collectors in each, or many, of the large fishing communities. It could also follow, that any strategy which falls outside these basic parameters for data collection should receive their funding from other than government sources, or, if these are at the request of the fishing industry, then the industry should bear the cost, especially if can afford to do so.

The initial data collection task is a complete census of fishers. One cost effective strategy suggested is to use inter-agency resources to collect this data when the regular census of the population is taken. This can assist in providing basic, and later, updated information where this is needed. The key role of the data collector, following the establishment of the data base, is the verification of the fish caught and that fish which is landed with additional information on the area of capture, size, age and sex of the catch. The updating of the initial data base can often be accomplished during regular data collection activities. Some states use personnel on a part-time basis for both data collection and enforcement, but this has conflict of interest difficulties where fishers may believe that the only reason personnel are collecting data is to "catch them" in non-compliant activities. This can lead to falsification of the data provided.

Many countries already have extension officers, and community development officers from one ministry or another and these personnel can provide a wealth of information if it is channelled, or accessible, to the fisheries database. Data collection reliability is often enhanced when the individual collecting the data is known and respected in the community. The data being collected are, after all, personal information on catches, fishing areas and income, all which can be interpreted as being sensitive to the fishers.

There are several options for the offshore portion of fisheries MCS. These depend on the value of the fisheries to both domestic and international fishers. Under the Convention on the Law of the Sea, the coastal State may establish the terms of fishing resources surplus to its current harvesting capacity. This includes a potential resource rent for the opportunity of fishing in the zone. This rent must be reasonable, but it can also be used to offset some of the costs of enforcement.

There may be several options to survey offshore fisheries activities. For example, it may be safer and more cost-effective to conduct inspections and transhipment of fish in port rather than at sea. Where possible, maximum use of port surveillance and inspection activities are encouraged as these can be very cost-effective and cheaper than at-sea inspections. It must be recognised that at-sea surveillance will still be a necessary component of most fisheries regimes.

Fisheries Management Scheme

One theoretical scenario of cost effectiveness could be a fisheries management scheme based strictly on effort and area controls, instead of quotas by species which wander all over the zone. The effort limitations would be established from catch rates in the past applied to each fishery. These could be verified through information provided on the landing of the fish. In this case, the surveillance aspect would be greatly facilitated by being based on vessel sightings and effort monitoring instead of hands-on inspection of the vessels at sea to determine their catches.

Using the principle of least cost to the State, however, if the fishing industry wishes to control the fishery by quota management instead of effort control, they might be encouraged, as part of the resource rent, to shoulder, or share the additional costs which may be incurred with such a system. Perhaps a combination of these principles could be effective if the fishing partner were willing to assume additional costs for surveillance to offset air coverage, observers or at-sea inspection costs for management schemes which they prefer to have implemented. This interactive management style would require a high degree of input and the acceptance of responsibility by the fishers for the conservation of the marine resources. In many cases it would require a complete change in attitude, consequently, it is a longer term strategy which one can consider for implementation with education of the next generation of fishers.

On the issue of quota controls, these have been found to be effective only if there is a timely acquisition of accurate catch data, including discarded and dumped fish that has been removed from the resource base. The acquisition of these data is also helpful for stock assessment, but it requires at-sea observations and inspections for quality control of the data being collected. This latter requirement is an expensive undertaking. It is, however, one rationale for fishers to assume the total costs of observer coverage, if quota control is the strategy they support for fisheries management.

Fisheries Legislation

Fisheries legislation forms a major component of the control aspect of fisheries MCS. The fisheries management plan is transferred from theoretical ideas to legal requirements which form the base for the MCS operations, through the drafting and passage of fisheries law. It is at this juncture that fisheries managers, MCS officials and lawyers can assess the enforceability and cost of their management schemes. Common concerns expressed by MCS personnel all over the world are that the written law is often untimely, it no longer reflects the full intent of the management plan, it is overly complex, unenforceable and consequently, expensive to attempt to implement. The latter impacts on the credibility of fisheries officials in the eyes of their clients, the fishers.

Cost effectiveness can be enhanced for this component if there is coordination and cooperation between the fisheries administration and legal drafters. This is an investment in time and greatly facilitates implementation of the final product for the benefit of the State. It also serves to strengthen the knowledge and capability of the legal authorities in fisheries resource management, a potential benefit to the development of other resource management legislation. It falls on the fisheries administration to foster this linkage with the justice department. The internal policy decisions in the control phase of fisheries management relate to the strategy for implementation of the MCS operations and will be different for each fishery and state.

Licenses

One of the most important tools of fisheries management, which is often overlooked, is the privilege to fish, the fisher's license. The license can be used to provide all the base data regarding fisheries activities in the zone. It can require appropriate reports on fishing gear, activities in terms of time, location and catches, and can also require cooperation in fisheries management objectives. Further, it is the main tool which will serve to obtain the resource rent for the privilege of fishing in the State's waters. It can be the tool which establishes controls on an otherwise uncontrolled, open access fishery to meet the State's obligations under the Convention while minimising the cost to the resource owners, the taxpayers. It can also be used to offset funding requirements for additional surveillance measures. The supporting legislation which makes the license so important is the legal right of the State to grant or remove this privilege and to exact sanctions for non-compliance. The license therefore contributes to the monitoring component of MCS through the requirement for base, and operational, data. It is the legal instrument for the control of the fishery and is key to the implementation of fisheries management plans.

Resource Rent

The principle of "keeping it simple" is a concept to keep in the foreground in the surveillance component of MCS. The current times of fiscal restraint make this even more applicable today. Using this line of thought, coupled with the idea that the fishers should assist in funding any diversion from bare necessities for fisheries management, the following are examples of strategies which may be considered. Assuming that fisheries management plans have been developed and legislated in a cost effective manner, the Fisheries Administrator must then look at the human and infrastructure resources necessary to conduct fisheries surveillance operations.

Assuming that resource projections have been determined based on one of the many stock assessment models, and that the basic requirement of the State is to ensure that the removals from the resources do not exceed these agreed totals, many options exist for enforcement. If there is only a requirement to report and land all fish captured, the surveillance function can possibly be carried out mainly with shore monitoring and spot checks at sea. It may also be the best strategy to use in a management system where the coastal resources are to be managed by the municipalities or sub-regions of the country, such as in Germany or the Philippines. Countries where fishing zones are very small and coastlines are short might be States where the economics of the fishery would be best served by such a policy, if landing facilities are available.

Other zones, where there are large offshore fisheries or extensive coastlines, may require a strategy which depends on offshore technology. Some examples include the offshore component of the Philippines, the extensive zone of FFA, Australia, UK, New Zealand, Namibia, Morocco and Angola, as well as the island states of the Indian Ocean and the Caribbean Basin. Surveillance technology comes in various expensive packages including radar, aircraft, and satellite technology and cost effectiveness in these instances is very important.

No Force Tools

The use of cost effective "no force" tools is becoming most popular. These include the use of national or regional registry systems where the threat of removal of "good standing" is often enough to ensure compliance. Unfortunately, there are no international conventions in force concerning registration of ships, and none at all being considered for fishing vessels. The concept of international vessel registration standards and exchange of information would greatly facilitate the identification of vessels, implementation of flag State and port State control mechanisms and establish controls on "international renegade fishers" who are conducting fishing operations which undermine the internationally respected principles of conservation. The potential for the use and expansion of the regional register into an international principle merits serious consideration.

Another, "no force" initiative is the flag State responsibility for the activities of vessels flying its flag. This has been used effectively in the FFA treaty with the tuna fleet of the United States and is the principle behind the FAO "flagging agreement" and the efforts to control the fishing of highly migratory species and straddling stocks. Another mechanism is the use of observers without enforcement powers, which, while being effective for data collection, has also been found to be a deterrent to non-compliant activities. This latter strategy is not without its pitfalls, the chief being the potential of external pressures on the observers at sea resulting in less than desirable data collection practices.

A new initiative is the development of vessel monitoring systems for timely catch and position information. The acceptance of this technology in the courts is one of the future challenges facing this technology. Finally, a growing trend, borrowed from the commercial vessel trade, is the development and use of "port State controls" whereby there would be international agreements struck on a regional basis for the inspection and enforcement of fisheries legislation on any vessels operating in the entire region. This is an effective, low cost control using the potential of any country in the region being able to detain non-compliant vessels and crews as a counter-incentive to non-compliance with respected international maritime principles, be they for fisheries, pollution control or safety-at-sea.

These technologies can all be cost effective, and where they can be applied appropriately, they can be of little cost to the State other than the investment of time for coordination.

Management Measures

Management measures are the specific elements of fisheries control which are embodied in regulations and which become a focus for surveillance activities. Cost-effectiveness needs to be considered for each management measure. The fisheries management plan, operational strategy and the management measures chosen for MCS must be included in the fisheries legislation. This will provide the base for implementation of the fisheries management plan.

One measure to be addressed is the use of mesh size for conservation purposes. It should be noted that this requirement can only be enforced in two methods; inspection prior to going to sea, with the provision that no other gear can be carried on board for that trip; or, inspection at sea which can only provide a snap shot of the fishing operations while the officer is onboard the vessel. The fisher may wish to prosecute a second fishery where a different mesh size is authorized and the earlier noted requirement then becomes an inconvenience.

In the case of towed nets, if the fisher is permitted to use strengthening ropes to keep nets together when full of fish, and top and bottom chaffers to protect the nets, and then trawls through weeds on the way to the fishing grounds, can small fish really escape from the net? This is further complicated by the fact that many nets are constructed with diamond mesh which has a tendency to close when pressure is applied during towing through the water. Some countries are now requiring square mesh design which stays open during trawling, except under extreme pressure. One can then question the advantages of using mesh size as a conservation tool in the case of towed gear, noting the difficulty of ensuring that it works properly.

The case for gill nets and entangling nets is different and the mesh size can be a significant conservation factor, but in these fisheries there will be little need to carry more than one type, or size of net, to sea; consequently, it will not inconvenience the fisher significantly to have the vessel inspected prior to fishing and ensure only one size net goes to sea for the trip. The method for measuring a net should be standard, at least by country, and acceptable to the courts where fisheries cases go to court. Assistance of the judiciary to establish these standards would be advisable. Common standards include:

a) measurement with a standard, graduated wedge or an implement with a standard width,

b) measurement of the net when wet as it would be while fishing,

c) measurement of the mesh stretched between opposite corners,

d) measurement of several adjacent meshes and averaging the result,

e) measurement in the middle of the net away from any strengthening ropes.

A second management measure is the use of chaffers and strengthening ropes. Chaffers are attachments to the bottom and top of towed fishing gear to prolong the life of the expensive nets by reducing wear and tear from rubbing on the sea-bed. These can be made of netting, for top side chaffers, or leather strips, for bottom chaffers. These attachments are common and necessary fixtures to the gear, but a supplementary result is that they block the mesh and therefore retain all fish caught in the net. It has been found that the method of attachment of this gear does have an impact on the possibility of fish escaping the net. The bottom chaffers are usually heavy twine or pieces of leather, especially for bottom trawls where the net is dragging on the sea-bed. These should only be attached at the tail of the net (the cod end). Topside chaffers come in many different shapes and sizes, the main criteria normally established for this attachment being that it is attached only to the cod end of the net and in such a manner that it does not overlap the mesh to restrict the normal openings of the mesh in the cod end.

Strengthening ropes are necessary attachments to hold the net together and prevent it from ripping open when it is hauled on deck with a full load of fish. These are ropes which should be attached along the main axis of the net and where attached across the net must be attached in a manner as to ensure they do not reduce the size of the meshes in the net.

Strengthening ropes and chaffers can only be checked during inspections in port and at sea. On larger vessels operating for days at sea, these can be re-laced to the net at sea and therefore become a management measure which is difficult to control without at-sea inspections or continual monitoring during the fishing operations. In the case of smaller vessels working inshore, port inspection would normally be sufficient as it is impractical to make changes during such a short trip.

Area closures are a management tool recommended to protect spawning areas during known seasons. This tool is also used to separate types of fishing gear whereby one type of gear is permitted, such as set gear, and another is prohibited, such as mobile towed gears. One lesson from the experience of area closures is that closures based on depth of water are unenforceable. More effective is the latitude and longitude designation of the area for such controls. This tool does not require at-sea boardings and inspections unless there is a prohibition of different types of stationary gear, such as traps and nets. The surveillance of area closures can be accomplished by properly equipped aircraft which have photographic and integrated navigational equipment and night flying and surveillance capability. This can minimise the necessity for sea patrols, but the presence of a patrol vessel is still the best deterrent. It has been noted that air patrols, although expensive, cover a vast area and are more cost effective in gaining necessary fisheries management and MCS information than sea patrols.

Some countries use windows as a tool to control fishing in their zone. In this case, areas for authorised fishing are established by latitude and longitude and all other areas are closed to all fishing. These windows can be set for different gears and patrolled by aircraft. Windows or area closures can be used effectively to reduce gear conflicts, especially between offshore commercial and artisanal fishers. On some occasions, countries have set area and time closures to permit different gear types to use the same area, but at different times. Another advantage of windows as a surveillance tool is that sea patrols can then be concentrated on these areas for at-sea inspection for compliance with regulations which are not enforceable by any other means.

The disadvantage of windows, is the impact this management strategy may have on the fishers and their reduced ability to chase the fish over their migratory paths. Windows can restrict fishing to the point that fishing opportunities are not viable. It is best to ensure that such a strategy does not overly impede the chance for fishers to access the resource. Stocks with low migratory patterns could be considered for this management measure.

There are various types of catch or quota controls used by governments today. These can be daily, seasonal or trip catches, zonal quotas, vessel quotas and annual quotas. In each case, there is a requirement to be able to verify, on a timely basis, the actual catches of fish by each vessel, species and possibly by area. This becomes a very complex administrative programme which is expensive in terms of finances and human resources. The monitoring and at-sea inspection and enforcement requirements to implement such a system are significant in terms of communication costs to support the necessary processing and analysis of data and verification of data quality. It is commonly accepted that control of the removal of resources from the sea is the most desirable conservation measure. Whether this is done through controls which count each fish or through less complex and less expensive methods, such as control of fishing effort, is a point which can significantly impact on the cost-effectiveness of fisheries management and MCS operations. It is accepted that where estimates of stock abundance are accurate, then catch controls would maximise the benefits to the fishers by permitting them the maximum removals, but this is an exercise which, to date, has been found to be very costly.

Trip limits are sometimes used as a measure of fishing control. These limits could include total catches permitted per trip or, more commonly, effort limits. The former requires someone to meet the vessel upon arrival to be present during the weight out of the catch. If a fisheries official is employed to monitor landings for the data system, this strategy would blend in with normal operations. If not, then it would necessitate extra personnel and effort. The disadvantage of trip, or catch, limitations is the temptation to discard all the lesser value, or small fish, prior to landing, thus possibly promoting the dumping of fish. The effort control requires reports of departure and return and a capability to verify these periods through sightings at sea. This is not impossible however, and will be discussed further as a viable, cost effective control mechanism.

A possible alternative to catch and quota controls is effort control. Where stock assessments are not very precise, there can be a measure of fish removals through effort controls by limiting the fishing time of vessels and fishers in certain areas. This can be implemented by assessing the experience of fishing units and then, based on a conservative assessment of the fishing efficiency of each fishing unit, limiting fishing to the time expected to take a certain amount of fish, with an appropriate safety factor for conservation. This can then be cost effectively checked through air surveillance and verified by port inspections prior to, and upon completion of, fishing in the zone, for the large offshore vessels. Coastal patrols and landing checks are usually sufficient for the artisanal fisheries. The at-sea resources then required can be minimised, thus making the system more cost effective. This is a system which can be implemented to establish some control of effort even where the exact resource status is not known, but where catches appear to be consistent in terms of quantity and size of fish. This would be a temporary measure until appropriate data gathering could be established and data analysed for more accurate stock assessments. There would also be the need to seek the support of the fishers for information on the current fishery and to gain their support for such controls as a temporary measure where the situation may appear stable, but is actually vulnerable, due to the lack of data.

A growing trend in many countries is to use individual transferable quotas (ITQs) as a management tool. This system is employed for those fisheries where there is limited access to the resource, the numbers and identification of each fisher and gear are known, recorded and have a history. The system then provides for an allocation of fish to each fishing enterprise, or fisher. This is often by species and stock area for a specific period of time. The fisher then has the right to plan the fishery of these allocations during the period to maximise economic benefits. The fisher also has the right to transfer, or sell, the allocation, or portion thereof, to another fisher on a temporary basis. This initiative enhances the benefits to the fishers with respect to their cost effective harvesting and processing of the resource to maximise their economic returns.

The complexity of the system and the requirement for an advanced communications and data network to effectively manage this strategy makes it difficult for developing countries. Experience has demonstrated that this strategy is only appropriate for small island States with small fleets and few fishers, fishing for essentially an export market. These prerequisites are necessary for States to be able to implement the appropriate controls to successfully use this management strategy. It is for these reasons that the ITQ strategy is not recommended for developing countries at this stage of its evolution.

The ITQ system does provide an opportunity for success if there are available resources to successfully use new technology. Such an initiative is being attempted by the FFA at the moment and the successful implementation and enforcement of the vessel monitoring system (VMS) could make this an ideal management tool for the future. The key to the success of the ITQ system could be the VMS system if its eventual cost to countries and fishers is low.

Another management tool is the establishment of minimum or maximum fish sizes. The regulation normally specifies that the capture or landing of a fish of a certain size is not permitted. The intent behind the minimal size of fish is to prevent the harvesting of noncommercial sized, juvenile fish of little market value. This is then expected to assist in stock enhancement, maximising future benefits to the fishers. The prohibition against large fish is usually intended to preserve the brood stock. Unfortunately, neither of these regulations can be enforced without continual monitoring at sea. The prohibition against landing can easily be subverted by culling the fish and dumping the prohibited sizes. Once the fish is caught, it is usually dead. Dumped fish are not normally recorded as catch; thus the estimates of removals from the stocks are therefore undermined. It is suggested that fishers be encouraged to land all their catches, that these be analysed against their areas of fishing and if areas need to be closed to protect fisheries, then this be done. This implies that the use of fish sizing would not be a legislated management tool, but an indicator for fisheries managers to close an area where small fish, or large brood stock, are being caught. This suggestion might bear further consideration, especially where fishers are fishing in one area where all fish must be retained and also in another where possession of under-sized fish is an infraction.

The prohibition of certain fish sizes, although it may encourage the fishers to shift fishing zones, also encourages dumping and misreporting. The surveillance costs to enforce these regulations are difficult to justify unless there is a physical presence on the vessel at all times, or a high level of spot checks at sea. Most states implementing a MCS system employ some sort of vessel movement controls. These are usually in the form of report requirements from the offshore foreign vessels and trip intention reports from the larger domestic vessels which are fishing longer than two or three days at sea. The vessel movement reports for the foreign vessels range from zone entry and exit reports, to port entry and exit reports and area changes.

In the case of domestic vessels, the port departure/entry and area change reports provide the basic movement information for MCS purposes. All movement reports require the vessel identification which includes the vessel name, call sign and the master's name as well as the fish onboard by species and intended activity. In the case of the foreign vessels, the fish onboard and intended activity are important for surveillance and catch monitoring in the zone. If the vessel has already picked up its fishing license at an earlier date, then the MCS authorities need to decide whether an inspection is necessary and if this will be at sea or if the vessel will be ordered to port. This first zone entry report for foreign vessels draws up the information for the vessel and commences the monitoring exercise, which will continue until the vessel has departed the zone and all reports and documentation have been received.

The zone exit report, which is commonly required by the central fisheries control, before departure from the zone, reiterates the vessel identification information, and the time and position of expected departure from the zone. This report is the final opportunity for the Fisheries Administrator to inspect the vessel by intercepting it at sea or ordering it to port. The matter of permitting multiple exits and entries to the zone for fishing purposes is one with which Fisheries Administrators will have to contend. A vessel could have good reason to exit and enter the zone due to medical reasons or such, or it could be for fishing operations outside the zone, transhipping its catch, changing fishing crews and thus avoiding coastal State regulatory requirements.

As all the latter reasons impact on the fishing efficiency of the vessel, the coastal State may wish to include in its legislation and fishing agreements, provisions to capture the information required for conservation purposes. If coastal State fisheries officials are aboard the vessel upon departure from the zone, their authority for fisheries surveillance and enforcement may be challenged. It is best in these sensitive situations to decide early whether the vessel will be permitted to depart the zone; if not, the order to "halt" must be given prior to departure from the zone to establish the parameters for "hot pursuit". The zone exit report thus triggers the action for decisions regarding the control of the vessel prior to its departure from the zone.

Vessel sightings reports are collected to update the fishing vessel monitoring information database. These reports are usually standardised and are completed by all MCS resources and sent to the MCS control centre for collating and updating of current information. The report normally includes vessel identification information, such as the name, vessel marking/call sign, home port and information as to the activity of the vessel. If the vessel is steaming, the report includes its course and speed, if these can be determined. If it is fishing, the course, speed and type of gear it is using are necessary. If photographs can be taken of the fishing activity, including the gear, e.g., lines in the water, etc. these should be taken and labelled with the vessel identification information, position and time of the pictures. The vessel sighting report should also record that photographs were taken, and the numbers on the film. This could all be necessary information for the courts if it were later found that the vessel was not fishing in compliance with its license, or fishing without a license. These are fairly common standard procedures which all MCS personnel should be trained to do on each sighting. Sightings form the key verification of the vessel fishing effort in the zone and can also be used to estimate catches.

Vessel inspections are a key management tool for monitoring and surveillance. Port inspections are considerably easier to conduct than those at sea due to the safety factor of not having to deal with the motion of the sea either on boarding and disembarking, or during the inspection itself. The detail of the vessel inspection, either at-sea or in port, depends on the Fisheries Administrator. Naturally, it will be impossible to see the fishing and processing operations during an in-port inspection, but it should be possible to reconstruct the fishing activities of the vessel since its entry into the fisheries waters of the State. In both cases, one should be able to determine the fishing pattern, catches, and verify the fish onboard the vessel through an inspection. It should also be possible to check the storage and size of the fishing gear, at least that on deck, for compliance with fisheries legislation.

The fish onboard the vessel can be determined as precisely as desired if the funds for off loading and reloading are available. This is not normally recommended unless there is sufficient reason to believe there has been a violation. The advantage of at-sea inspections is the monitoring of the handling of the gear, the processing and storage, and verification of the processes for handling waste fish. This can provide information for fisheries management planning and possibly catch and effort information which would be useful in access negotiations with foreign partners. Efficiencies of processing methods and equipment can be monitored and used to validate conversion factors from processed to round weights.

The accuracy of vessel inspections, both in port and at sea, is a crucial component of the surveillance aspect of MCS. It is the initial inspection which verifies the fish onboard the vessel and forms the base data from which final assessment of fish caught in the zone can be determined. The intermediate inspections during the authorised fishing period provide

local operating conditions and make the assessment and decision based on these results, with appropriate attention to the cost of procurement and, more important, operations and maintenance costs and the capability to carry these out in the country. New suppliers, if they wish to sell this equipment, should be willing to fund trials in the local environment if they are serious regarding the efficiency and effectiveness of their equipment for the task to be done. It is the supplier who should shoulder the financial responsibility of assessing the appropriateness of the equipment to the situation before attempting to sell this equipment. It remains the responsibility of the Fisheries Administrator to assess the performance. The operations and maintenance training for the equipment should be included in the cost of the equipment and provided by the supplier, if it is accepted.

There has been a growing trend, due to the relatively high cost of MCS operations, to seek "no force" MCS strategies. The intent here is to pick those MCS tools which can exercise sufficient monitoring and surveillance controls over the fisheries resources to meet government needs at the lowest cost possible. The most popular "no force" strategies seem to be the use of a national, or regional register of vessels of good standing. Information included in the register includes the identification of the vessel and master and a record of performance and compliance in the fishing sector. In a regional situation, information is shared between parties and decisions can then be made regarding a vessel and master in the entire area. Vessel masters and fishing companies have, over the years of experience in the FFA, been seen to respect this tool, to the point that potential removal of good standing has been enough to ensure compliance or appropriate action in cases of a difference of opinion regarding fisheries MCS activities.

Vessel registration is an area which merits considerable attention in the future for both fisheries control/MCS purposes and also vessel safety requirements which impact on safety-at-sea and protection of the fisheries habitat. At present there are no international conventions or standards for registration of fishing vessels. The potential exists for this to be a credible international management tool for both flag State and port State control mechanisms which could benefit developing countries with information on third party vessels applying for registration in their countries.

Other tools in this category of "no force" mechanisms include the port inspections, flag State responsibility for the actions of its vessels in the zone and observer programmes. Many countries have required a representative from the international fleets which are authorised to fish in their waters to be resident in the State and to be responsible for the actions of the fleet in these waters. This has been found to be effective only if the representative has the appropriate authority over the vessels in the fishing fleet. New tools in this category could include the idea of effort control, vessel monitoring systems and others which would require minimal effort and expenditure from the coastal State for use of its MCS resources.

Fisheries Management Plans

The development of fisheries management plans is, ideally, the result of analyses of biological, social and economic information and appropriate formulation of fisheries management strategies. However, fisheries management needs a legal base from which to implement fisheries operations for the conservative management of fish resources. The legal instruments which form this key legislative base include the fisheries acts and regulations developed pursuant to the planning exercises. These legal documents must, of necessity, for national and international credibility, take into consideration the terms of the Convention on the Law of the Sea, and any bilateral or multilateral treaties or agreements in effect regarding fisheries in the zone of influence and control of a given country.

Both bilateral and multilateral agreements can have a significant effect on fisheries management and MCS systems and strategies. Negotiation for access to fishing grounds of developing countries can be an interesting and very challenging exercise. The coastal State is looking for financial benefits from the negotiations, whereas the international party is looking for access to the fishing zone. This usually results in compromise which in effect will undoubtedly impact on the MCS strategy and the legislation for fisheries management implementation. The fact that it is usually a distant water fishing nation which is seeking fisheries access, possibly having greater experience in the process, with an established pattern for negotiating benefits weighed heavily on their side, can place the developing country in a very disadvantageous position. The coastal State's requirement for foreign currency can also place undue pressures on the negotiating team in favour of the negotiating strategy of the foreign party.

A review of past agreements of distant water fishing nations has demonstrated the advantages to these nations, sometimes at the expense of the developing countries. Negotiated international legal agreements take precedence over existing legislation and consequently, fishing vessel licenses for these fleets sometimes have clauses which give them a fishing advantage over other fishers. This is not to say that international fisheries agreements are not beneficial to coastal States, as many countries have received considerable assistance through these processes. Whether states have received full value for the resources they have negotiated away is a matter for conjecture.

An example of such negotiations concerns the EEC, especially since expansion of the fishing fleet with Spain and Portugal joining the Community, whereby their aid package was very much linked to fisheries negotiations. For several years, the EEC provided fisheries aid and training for all fisheries components except MCS activities, and there was no explicit recognition of the rights of artisanal fishermen in their agreements. Recent agreements with the Seychelles and Madagascar have been more comprehensive but the developing State often does not have the capability and infrastructure to implement the letter of the agreement.

Other distant water fishing nations have refused to recognise regional fisheries agencies and have insisted on bilateral negotiations. This tactic has now been all but broken by the FFA, through the internal regional agreement to insist on minimum terms and conditions for international fisheries agreements. This accomplishment is another achievement of regional cooperation.

One of the common concerns of fisheries personnel lies in the final output of litigation processes. Fisheries Administrators have expressed their concern over the apparent lack of success experienced in this area of MCS operations. There is definitely the need for training in this activity to ensure that evidence is gathered correctly and presented in such a manner to result in successful prosecution of fisheries cases. Another influential factor is the legal system of the States. The system to gain a conviction for an offence in the civil law states differs from the system in the common law states. While in the civil law system, the administrative procedure for imposing a sanction provides for an appeal before the Minister, this is not always the case in the common law system.

In relation to evidence under the common law system, certain technical rules, such as the rule against heresay, restrict the range of evidentiary material that can be introduced. This can cause difficulties for the introduction of certain types of information in the courts, such as that obtained from vessel monitoring systems and legislation may be necessary to accommodate this type of information. On the other hand, in civil law states, a report from an officer is accepted as prima facie evidence. Without going into significant detail on the differences between the two systems, this provides and example of the impact of the legal system on the implementation of MCS strategies.

Fisheries management needs a clarification of the authorities of the ministry responsible for fisheries and its powers over the legislation and implementation of these laws in the offshore, coastal, riverine and lake waters. The authority of the Minister is required with respect to powers to enact legislation through regulations to implement fisheries management plans. Here the civil and common laws differ, in civil law the Director General for Fisheries has the authority to issue licenses for fishing activities, cancel or suspend said licenses, appoint persons to positions of responsibility in the Ministry, and to exact administrative penalties for fisheries offenses.

In the common law system this authority rests with the Minister, who may delegate this authority. In the case of civil law there is an appeal tribunal usually available to the accused where a serious offence has been committed while in the common law system this option is not always present. The specific authorities and powers of each key member of the fisheries department, management and field staff, are normally amplified in this section with respect to management authority and punitive responsibility and powers, monitoring and data collection requirements, sampling, inspection, search, seizure, detention, prosecution and confiscation procedures. Further, the law usually formally states the appropriate reception expected for fisheries personnel and observers or others acting under the authority of the government. This latter point is very important for governments which may consider privatising aspects of their fisheries MCS activities and for certain aspects of monitoring, such as dockside monitoring and observers.

The responsibilities, as well as the authorities, of each of the individual positions is also stated in the legislation. An example is the responsibility for fishers to accommodate and feed observers at the officer level of their vessels, but the law could also note the monitoring and advisory role of observers excluding enforcement powers. The details of the fisheries management plans are outlined in the law and the obligations of fishers in this respect are then stated. The heavy pressures on most fishing stocks have resulted in the consideration of more stringent conservation measures including not only measures that regulate fishing activity, such as closed seasons and areas and gear types and uses and also, fishing effort. This latter point can change fisheries from open employment opportunities of last resort to a closed fisheries profession through the imposition of a limited entry fishery concept. The terms of entry, the requirements for licensing and criteria for application for such licenses can also form a part of the legislation.

With respect to the use of licensing as one of the management tools, the two tiered licensing system (one license for the vessel and gear and another for each fisher) has been most popular and effective to date. This is opposed to licenses for each component, vessel, gear and fisher, or the combination of one license for the fisher which is further limited by a particular vessel and a specific gear type. The license application forms collect the basic information for the MCS database for both fisheries planning and MCS operations. Information on the fisher, vessel, gear, operations planned, social aspects and demography is also collected during this process.

The obligations and requirements for certain management measures are set under the fisher's licenses. They include the requirement for marking of the fishing vessel for ease of identification. This requirement has now been standardised throughout the world as a result of FAO efforts in 1989. Markings are clearly specified as to the size and location on the vessel, reflecting the size and type of vessel. The vessel marking requirements are now recommended for inclusion in all fisheries legislation. This facilities both the monitoring and surveillance aspects of MCS operations from the air and the sea.

An increasingly common requirement for fishers is to mark their fishing gear. This has been assisted by the standard definition and classification of fishing gear types which should be a document in the library of all Fisheries Administrators. The requirement to mark the fishing gear in a clear and obvious location will also facilitate the monitoring and surveillance aspects of MCS activities. The matter of logbooks has been one of considerable discussion for several years. Logbooks serve many uses, first for information on the fishing operations of the vessel and, second, this information, if in sufficient detail, provides important input into the biological stock assessment component of fisheries management. It is these two aspects of logbooks which create discussion between fisheries managers and scientists, the latter seeking additional detail and knowledge.

The MCS requirement for logbooks is for monitoring of the levels and the areas of the catches, which then contributes to future fisheries management planning and also serves as a verification of compliance with the terms of the fishing license. The biological component requires greater detail on the fishing operations which may include, in addition to the aforementioned, the depth of fishing for each set, or haul, length of the gear and/or time of trawling, species breakdown by set, or haul, size of fish and age and temperature of the waters during the set, or haul, at the depth of fishing.

The economist would also like to know the processing methods, packaging and labelling, waste factors, and processing of fish offal. Other fishing operations such as transhipment of fish, quantities by species and location and names of the vessels involved in the transfer are also details MCS officials wish to record and analyse. These points are presented to note the complexity of the discussions regarding this topic in the past. This has lead in some situations to two or three logbooks for vessels for fishing, production, and transhipment. In addition to the requirement for this logbook information, it should be remembered that this is one component which is commonly standardised for regional cooperation, if such an initiative is being contemplated. It is obviously necessary to agree on some standards for the collection and recording of the data in a manner which is usable to fisheries staff.

Most countries have found it effective to design their own logs and issue them, with instructions, to fishers. It has proven to be less than effective if information that is not needed is collected, and this should be avoided, regardless of the temptation to gather it. Transhipment has been an issue of some importance to MCS operations. There are several philosophies regarding transhipment, ranging from no permission for transhipment inside the zone to permission for transhipment in the zone, but with prior notification to authorities, and finally, to permission for transhipment, but only in port.

Fishing vessels normally resist transhipment in port, due to the bureaucracy involved, as they can be treated in the same manner as normal transport vessels with no recognition that they are in a special category, due to their cargo. When there is no transhipment allowed in the zone, this will inconvenience the fishers, but the impact is also a loss of data on catches taken in the waters of the State, as the vessels will tranship outside the zone where there is usually no requirement for the master of the vessel to permit an inspection. Verification of the catches in the zone will be lost, unless there is a requirement to report the fish onboard on entry and if exit from the zone is not permitted without an inspection.

This becomes administratively difficult. Transhipment inside the zone is difficult to monitor and has its level of difficulty with respect to safety-at-sea. The recommended approach is the requirement for all transhipment of fish to be authorised for coastal State ports recognising that although this may be expensive for the vessel, and hence impact on the revenues expected from the access agreement, it must be balanced against the risks of inaccurate, or inadequate, information on retained catch. The monitoring aspect of port transhipment is safer and much more accurate, thus being a benefit to MCS activities. It is for this reason that countries may wish to review their port State administrative requirements with a view to encourage fishing vessels to tranship their catches in port. The legislation must reflect the decision in this regard.

If the management strategy is based on effort control using assumed daily catch rates, including a factor for culling, there would then only be a need for a report of the vessel identification and its positions of fishing over the period. The requirement for the return of completed logbooks and a port inspection prior to leaving the zone can be effective methods to verify retained landings, and possibly catches, for future effort control measures as required. This strategy would be strengthened by the legal obligation to tranship fish caught in the zone in local ports. There would then be a requirement for port inspectors, air surveillance and possibly fewer random sea inspections.

The use of satellite technology with vessel monitoring equipment could also be effective in this case. The cost effectiveness of reduced sea inspections could be significant. It must be noted that this strategy would only have potential for single species fisheries and could not effectively accommodate multispecies and mixed species fisheries. The requirement for periodic position reports from all fishing vessels during their fishing activities has become a standard practice. The rationale for this is twofold; one, for verification of compliance with legislation and licensing provisions, and two, for input into research programs on stock assessment.

Many countries require the vessel to report prior to departure from the zone and the intended departure point. This then permits planning for post harvesting inspections at sea if this is a management decision. Other management strategies follow the domestic principle whereby the vessels are required to come to port on completion of fishing in the zone. This latter serves many purposes. On the domestic side, certification of the fish caught and retained is easily obtained from weigh outs of the fish off loaded. This assumes there has been no transhipment of fish at sea, an operation difficult to monitor without regular air surveillance. With regard to international fisheries, the final port visit permits verification, through estimates, of the fish onboard and, taking into consideration authorised fish transhipment, will provide an estimate of the fish caught in the zone as a check on the reports from the vessel and later, the logbooks.

Towards Successful MCS Operations

Success of fisheries MCS operations in almost all cases has been due to the liaison between parties with a vested interest in the fishing industry. Clear delineation and acceptance of roles, responsibilities and obligations has made the process easier for all concerned. The harvesting of any natural resource by strong, independent business persons has always been a challenge and nowhere is this more evident than in fisheries. Successful MCS operations have reflected the absolute requirement for all participants and individuals with input into the process to understand and accept the management plans.

Inputs of Fishing industry

The purpose of rational, sustainable fisheries management, and subsequent MCS activities in support of this concept, is for the sole purpose of providing continuing benefits to the fishers of current and future generations. It would seem apparent then, that key to any discussions regarding the fishing industry would be the fishers themselves. Unfortunately, there is often little input from these, the grassroots clients and beneficiaries of the fishery. It is no surprise that, after many years of neglect and little input into fisheries management fishers are suspicious of government officials. Fishers prosecute the resource for financial gain and they do this armed with the combined knowledge of generations of practical knowledge passed down through families and the folklore in the communities. There is a need for respect and partnership of government officials and the fishers themselves in the development and implementation of fisheries management plans in the future.

This linkage with the fishers can often be assisted through liaison with fisher groups, unions, or cooperatives, where these are in existence. Where these are not in place the community organisations themselves can serve as a good link to the fishers and their families. When the fishers become confident that the intent of the government is really in their best interests, the government will have a very influential and powerful ally in implementing its plans.

Fishers who are organised appear to be more successful in ensuring their input into fisheries management plans. Most governments have extension officers of some departments to liaise with communities. These individuals can greatly assist in this process of establishing links with the communities for fisheries matters. Trust and cooperation are key to this process. Government initiatives to assist fishers in communities to organise themselves for fish trade purposes may also be viewed positively and contribute to this overall exercise of fisheries management.

The input of bigger fishing enterprises with contacts on the international market for fishing vessels, joint agreements and trade are also very important in this exercise. There is a caution from years of practical experience, however, in placing too great an importance on one sector of the fishery, while neglecting the impact other sectors have on the industry. The larger fishing enterprises should be encouraged to provide their input, and in fact will probably do so without urging due to their investment in the industry. The challenge will be determining the balance between interests of large fishing industry interests and the more scattered local artisanal fishers.

Inter-ministerial Liaison

The large number of ministries interested in the ocean sector and particularly fisheries creates the potential of administrative complexity. A listing of these ministries may assist in focusing on their respective roles and input that the Fisheries Administrator can expect from government partners. These include the Ministry of Justice with its judiciary, municipal and federal police agencies, port authorities, national defence, customs, immigration, health, foreign affairs and the fisheries department itself.

If one breaks down the interests, it can assist the Fisheries Administrator in the determination of the points of contact and timing for each party in the MCS process. The Ministry of Justice and the judiciary need to understand the fisheries management objectives, policies and importance of the resources and habitat to the State. Second, they need to have a good understanding of the intent behind the fisheries management plans and the MCS procedures to reflect these in the fisheries legislation. It is worth fostering the relationship between fisheries and the judiciary to obtain their assistance with fisheries legislation and its implementation.

The cooperation of the federal and municipal police agencies is an essential component of MCS. These agencies are often the operating arm of the surveillance component. Whether or not this is so, the cooperation of all enforcement agencies through shared databases and shared resources has proven very cost effective in the past. Where priorities can be arranged to coincide, it is always effective to join forces for enforcement purposes. Another consideration for fisheries with respect to other enforcement agencies is a delegation of powers and authorities for enforcement to the field personnel of these agencies. There is an obligation to the fishers resulting from this possible delegation of powers, however, and that is the assurance of appropriate training in fisheries enforcement, laws and procedures for these enforcement officials. The federal and municipal police agencies have contacts which can also be very useful in establishing the appropriate links with fishers to involve them in fisheries management and implementation planning exercises.

The Port Authorities can be of considerable assistance in fisheries MCS, both for monitoring and surveillance activities. Port State control is becoming a more popular initiative and a multidisciplinary port authority can facilitate coastal State port inspections, briefings, and transhipment of fish with a minimum of bureaucracy and maximum of control at low costs. The Port Authority can also assist in obtaining information regarding the vessel and its operations and can benefit from the sharing of the fisheries database with respect to vessel movement in the zone. Sharing of resources for monitoring and enforcement purposes could also be considered here as being potentially cost effective.

The Ministry of National Defence is very interested in fisheries operations under its sovereignty and State security mandates. There have been cases where military resources, having been seconded to fisheries and under fisheries control, have proven effective for MCS activities. The United Kingdom fisheries services operated in this fashion for several years in

England and Wales. In Scotland an executive agency under the Department of Agriculture and Fisheries in Scotland (DAFS) operates the government owned patrol and research fleet. This latter option has been found to be cost effective.

It must be stressed however, that MCS for fisheries is a civil police action, consequently, the use of the military as the executive power of the State is not really appropriate for direct involvement in this task. The military can however, in the course of their regular duties, assist in fisheries matters. The national defence departments of Australia and New Zealand have proven to be very effective in their air surveillance role for FFA over the years and have become an integral component of their system. On a smaller scale, sub-regional or regional organisations may be able to secure resources for similar services, or purchase their own aircraft and use seconded military personnel for flights and maintenance.

The sea component of the military has not yet been found to operate efficiently or cost effectively in a support role for fisheries in most MCS situations, due to the cost and bureaucracy of these heavily manned military resources. These high level military resources have been used by states in the apprehension of foreign vessels, but it has been expensive and control of the operation has been lost to fisheries personnel in this "civil police action". Only secondary tasking of resources from the military should be contemplated for fisheries MCS activities. The use of the military sea component is ineffective and inappropriate for MCS activities respecting domestic vessels and is difficult for foreign vessels. The sharing of the fisheries database with the military might also assist them in their priority sovereignty mandate.

Customs and Immigration are being considered together due to their similar interests of control of goods and persons into, and out of, the State. Again cooperation for surveillance services and sharing of resources can be cost effective, but only when priorities coincide. The American situation, where the U.S. Coast Guard enforces customs, immigration and fisheries legislation, creates difficulties, as the drug situation in the U.S. demands the resources on a priority basis. Fisheries thus have a low enforcement profile during these periods of drug interdiction. This drug interdiction requirement also influences the design of the vessels, which are usually inappropriate for fisheries where sea-keeping is a priority. One advantage of cooperation with these agencies is the sharing of surveillance information and secondary tasking for possible back-up support for surveillance activities. The sea aspects of these two departments are not complementary to fisheries and would most likely place fisheries personnel in a more hostile environment than would be appropriate for their normal function. Both of these agencies will also have an interest in fishing vessels during their port visits. Coordination with fisheries could facilitate these operations for the benefit of all parties.

The Ministry of Health usually has an interest in fishing vessels and the fishery for two main reasons; first, for international fishers, the state of health of their vessels and crew when they enter port and, second, for the standard of the product that is landed for domestic consumption and possible export. The Fisheries Administrator can gain assistance from these officials regarding fish product inspection concerns by cooperating on these matters.

Foreign Affairs has a considerable interest in fisheries, especially where it involves international fishing partners and negotiations. It is usually foreign affairs which takes a lead role in these negotiations and the challenge for Fisheries Administrators, as noted earlier, is to ensure that the bottom line for fisheries management and MCS control is not compromised in negotiations. Key to the negotiations is the principle of lowest cost, effective MCS to conserve the stocks. Any negotiated change that increases the costs of MCS should result only if there is a corresponding increase in benefits, preferably financial, from the international party to offset the increased costs for implementing MCS. An example is the case where the international fishing partner wants to have only one vessel with the fleet commander come to port to pick up the licenses for the full fleet.

The coastal State thereby loses the opportunity to verify the fish onboard each vessel for later calculation of the total fish caught in the zone. The offsetting agreement could be that the fleet commander's vessel would carry observers, paid by the international fleet, to the other vessels, where these representatives would estimate fish onboard prior to the vessels commencing fishing in the zone. An alternative would be an agreement for a fisheries officer to be transported by the fleet vessel to each other vessel, and then returned to port. A final alternative in this case could be an agreement for a grant of the equivalent funding to permit the patrol vessel to deliver the licenses to the vessels. In all cases, the vessels should not fish until the fish onboard have been verified by a fisheries official, or representative. These alternatives preserve the principle and importance of the first verification of fish onboard the vessel prior to its operations in the country's zone.

It is important that each of the MCS activities is clearly understood by the foreign affairs negotiator. There may be a tendency on the part of the international partner to attempt to negotiate other options which decrease the effectiveness of fisheries MCS. The Fisheries Units themselves, especially in the remoter areas of the country, need to be kept abreast of the fisheries MCS strategy. The heart of successful MCS operations in the field is the communications system. A telecommunications network is almost essential for MCS operations in the current fishing environment. Lack of information creates insecurity, concern and difficulties in supporting government policies and explaining these to the fishers. It is necessary for a credible relationship with the fishers that all the field units are fully aware of the policies and the rationale behind these decisions. On the other hand, it is also important that fisheries officials in headquarters also realise the benefits of the information which can be provided by their field units for all aspects of fisheries management and MCS operations.

Implementing MCS Plans

MCS plays key roles in monitoring and data collection, forming the legislative base for successful implementation and is the action arm of the policy. This makes it apparent that MCS must be an integral part of planning from the very commencement of fisheries management planning. Once it is determined which control mechanisms will be utilised for fisheries management, then the MCS officials can assist in planning an enforceable fisheries management plan for both the domestic and international fleets. MCS plans should be developed consecutively for each fishery, or group of fisheries. This will permit analysis of the legislation required and its enforceability. The assessment of MCS resources to be deployed can also be determined and all parties then can realise the pressures being placed on this infrastructure.

The international fishing plan, if international fishing is to be permitted, can follow a similar exercise, with one major additional input, the results of negotiations. Many factors for the negotiations have already been noted, the most important being the maintenance of control of the fishery in a non-discriminatory manner for both fishing fleets, national and international. The rule of keeping MCS simple and exacting benefits and remuneration for any divergence from this principle to recover the additional costs to the State is one to keep in the fore during negotiations. It is strongly recommended that senior MCS personnel be included in the fisheries negotiations to assist the chief negotiator in the technical aspects of the MCS portion of the agreement.

This can prevent difficulties in enforcement and misunderstandings in the implementation of the agreement. The impact of the negotiations then needs to be reflected in the fisheries legislation and in the MCS plan. The analysis and determination of the resource requirements and strategies to implement the agreement with an acceptable assurance of compliance is the following step in this process. Senior fisheries managers should then look at the total picture for MCS requirements for both the domestic and international fisheries and establish priorities on the use of the resources to permit MCS annual planning.

In this manner, all senior fisheries decision makers, up to and including the Minister, can be briefed by the Fisheries Administrator as to the expected deployment of MCS resources and the assessment of risk and consequences of the implementation plan. It must be accepted that there will not be enough MCS resources to fully address all aspects of the implementation of the fisheries management plans.

Operational Procedures

Some actual processes and core points for coverage in MCS operations, which have been used with some success in various parts of the world. The following are the components of operation procedures:.

Data Collection

In the case of data collection, a major component of the monitoring aspect of MCS, there are two preliminary and one follow-up activity in which fisheries officials become involved. They are the fishers licensing information, the vessel registration system, and the catch or effort monitoring system. The fishers and vessel monitoring databases, whether manual or computerised, will necessitate an initial census of fishers in the country and specific information on international fishers.

Vessel registration is intended to collect data which can be cross-linked to the licensing system, such as the description and size of the vessel, home port, call sign, where fish are landed, catch capacity in terms of hold and fishing gear type and capability, experience and efficiency as a fishing unit, the age of the vessel, outfit including communications, navigation and fishing gear, and processing capabilities, if any. The potential of regional and other international cooperation on developing and implementing standards for vessel registration have already been noted.

Fisheries economists and sociologists can use the information to determine the importance of the fishery to the national and community economy. The sociological profile of the fishers and their communities can also assist with enhancement of their position in the social and economic scale and provide support facilities where needed. Infrastructure, communications and data networks, is needed to support these data collection activities.

Fisheries Patrols

Fisheries patrols can be made more cost effective if planned with the view of integrating the surveillance resources to achieve the best results. All patrols should commence with pre-planning, a briefing of key participants to ensure that there are no surprises, the actual patrol, and a debriefing on completion, with appropriate documentation for record purposes, or follow-up action as required. Land patrols up rivers and along lakes and the coast can be effective if focused on fishing activity, areas of illegal activity of zones where the fisheries resources are particularly vulnerable to over exploitation by licensed and non-licensed fishers. Coastal areas, where domestic fishers operate and can be seen from land, can be watched for incursions by larger vessels not authorised to be in the area. This information can be relayed to coastal sea resources for action as appropriate.

Air patrols are most effective if the fishing areas denoted by season and species are known in advance. The air surveillance capability must be taken into consideration to determine the areas of priority, as endurance time of the aircraft will determine how many priorities can be addressed in a single patrol. The aircraft crew should be briefed on the patrol area and the expected activity in the zone, as well as the priority activities for surveillance. The air crew should also be provided with a summary of the vessels, their markings and authorised activities which can be expected in the patrol zone.

Random patrols, without a focus, have been found to be less cost effective than directed patrols for a specific purpose. Another priority might be an area of fishing concentration where local fishers have noted incursions of offshore vessels into their zones at night with resultant gear destruction.

Areas of heavy fishing concentrations where non-licensed vessels may hide during fishing operations could also be a priority. If stocks migrate close to the edge of, or beyond the fisheries waters of the State, there can be a temptation of offshore vessels to follow the fish into the zone if they believe the risk of the activity is small and they will not be apprehended. Air surveillance provides the front line information for the deployment of other more expensive resources, such as offshore patrol vessels.

Coastal patrols at sea are most effective if smaller patrol vessels can be pre-deployed to areas of fishing concentrations and operate from a base in this area to provide a timely response to fisheries conservation needs. In this manner, the patrol vessels can shadow the coastal fishing fleet for data gathering and verification and surveillance. The presence of a fisheries patrol vessel can also contribute to fisher safety, but this can be abused. There have been cases where fishers took turns to raise safety concerns to get a tow from the patrol vessel, thus putting the latter out of the patrol zone for a period while all the remaining fishers prosecuted the fishery in a spawning area.

Offshore patrol vessels, if it is decided that these are to be utilised, are best deployed to areas of concentrations of offshore fishing. Air surveillance is the primary tool which can detect area violations and unlicensed fishing activity. The patrol vessel can then be called to the scene, if necessary. More cost effective is the possibility of using diplomatic channels to bring the vessel to port, but this may not always be possible without greater international pressure than that which a single state can exercise.

Boardings

The decision as to whether it is safe to board due to weather is that of the master of the patrol vessel. The fisheries officer is the leader of the boarding team and as such it is the officer's decision whether the boarding party will actually board a particular vessel. On fisheries patrols there should be no doubt that the vessel is for support of the fisheries activity and hence the fisheries officer is in operational command of the patrol. The fisheries officer must ensure that the master and boarding team are briefed on the boarding procedures and expected support and communications for the boarding. The fisheries officer should ensure that the team is aware of the latest data on the vessel to be boarded, its license and fishing capability, design of the vessel and, if known, the route to the bridge of the vessel. The fisheries officer should have the latest pre-patrol data on the catches and last reports from the vessel for verification with onboard records. The fisheries officer than checks that the boarding team is properly equipped with documents, inspection equipment, such as gear measuring devices, safety and communications equipment.

Many fisheries officers make up their own package, in addition to that issued, to assist in the facilitation of the inspection. Some states have formalised these into quick reference field manuals for the officers. They may contain information on the fisheries regulations, commercial fish identification, fishing gear, common phrases translated for use in questioning the vessel master, special vessel identification markers for each country active in the zone, check lists, communications information and signals for use in special situations, measurement conversion graphs, and others. The fisheries officer should be aware if other fisheries department officials are aboard and what their assigned tasks are.

The boarding brief should delegate the activities of each member of the team for the duration of the boarding and any special instructions in the case of hostilities or resistance. If hostilities break out during the boarding, each team member should be aware of the disembarking procedures. The communications link to the patrol vessel is essential if there is potential for a less than friendly reception of the boarding crew. There are cases when the boarding team consists of the fisheries officer and an assistant, but these are rare, a minimum of four persons should be in the boarding team.

Inspection procedures

An inspection focuses on data gathering for two purposes. The first is for surveillance of the fishing operations to determine compliance with the terms of the license and legislation, and the second is to gather data for the monitoring aspect of MCS and fisheries management. The verification of the logbooks should be sufficient to reconstruct the fishing activities of the vessel since entry into the jurisdiction of the State. Most countries design their own boarding format to meet their data requirements and to facilitate computer entry, if available, and data cross-checking with other reports. Almost all reports have a section to identify the vessel, its license, confirm the name of the master, verify its activities over the period from entry into the zone, or from the last inspection. In the case of the latter, the report of the inspection should also be present.

Fisheries officers usually commence their inspections with a check of the licensed activity and verify that the positions and activities in the navigation/ship's logbook confirm that the path of the vessel is in compliance with the fishing license. Any reports made from the vessel to the fisheries department are noted and checked against their source. It is sometimes advantageous if the fishing log and catches, when summarised by the fisheries officer for the inspection form, are broken down into the same periods as that required by the State for the vessel reporting to the department. This facilitates cross-checking on arrival in port. The transhipment of fish by species and date is noted as well as the name of the vessel receiving the fish.

References

Allain, R.J. 1988. Monitoring, control and surveillance in selected West African coastal states with recommendations for possible ICOD interventions in the future. Ottawa, ICOD.

Bergin A. 1988. Fisheries surveillance in the South Pacific. *Ocean and Coastal Management*, 1988, Vol.11, No.6. pp. 467–491. Australia.

Clark, J.R. 1992. Integrated management of coastal zones. *FAO Fisheries Technical Paper* 327. Rome, FAO.

FAO. 1992. Monitoring, control and surveillance of fisheries in the exclusive economic zones of Asean countries: project findings and recommendations, *MCS training, completed manuals and course material*s. FAO FI:DPRAS/86/115. Rome/ Jakarta, FAO.

Newton, C.H. 1989. Monitoring, control and surveillance of fisheries in exclusive economic zones. *RSC Series* No. 49. pp. 209-213. Rome, FAO.

5

Impact of Climate Change on Capture Fisheries

Fish is highly nutritious, so even small quantities can improve people's diets. They can provide vital nutrients absent in typical starchy staples which dominate poor people's diets. Fish provides about 20 percent of animal protein intake in 127 developing countries and this can reach 90 percent in Small Island Developing States (SIDS) or coastal areas. Although aquaculture has been contributing an increasingly significant proportion of fish over recent decades, approximately two-thirds of fish are still caught in capture fisheries. Fisheries can also contribute indirectly to food security by providing revenue for food-deficient countries to purchase food. Fish exports from low-income, food-deficient countries is equivalent to 50 percent of the cost of their food imports.

Fisheries' Contribution to Economic Development

The number of people directly employed in fisheries and aquaculture is conservatively estimated at 43.5 million, of which over 90 percent are small-scale fishers. In addition to those directly employed in fishing, there are "forward linkages" to other economic activities generated by the supply of fish (trade, processing, transport, retail, etc.) and "backward linkages" to supporting activities (boat building, net making, engine manufacture and repair, supply of services to fishermen and fuel to fishing boats, etc.). Taking into account these other activities, over 200 million people are thought to be dependent on small-scale fishing in developing countries, in addition to millions for whom fisheries provide a supplemental income. Fisheries are often available in remote and rural areas where other economic activities are limited and can thus be important engines for economic growth and livelihoods in rural areas with few other economic activities. Some fishers are specialized and rely entirely on fisheries for their livelihood, while for many others, especially in inland fisheries and developing countries, fisheries form part of a diversified livelihood strategy. Fisheries may serve as a "safety net" to landless poor or in the event of other livelihoods failing.

Many small-scale fisher folk live in poverty, often understood as resulting from degradation of resources and/or from the safety net function of fisheries' for the poorest in society. This generalised understanding of the economic poverty of fishers in the developing world captures some of the situation of small scale fishers, but misses both the fact that they may earn more than peers in their communities and that their poverty is multidimensional and related to their vulnerability to a variety of stressors including HIV/AIDS, political marginalization and poor access to central services and healthcare. Small-scale fisheries, and especially inland fisheries, have also often been marginalized and poorly recognized in terms of contribution to food security and poverty reduction.

Exposure and Sensitivity of Fisheries to Climate Change

Climate change impacts on fisheries will occur in the context of, and interact with existing drivers, trends and status of fisheries.

Following rapid increases in production since the 1950s, the yield of global fish has stagnated and may be declining. Many stocks have been, or are at risk of being, overexploited. Statistics from the Food and Agriculture Organization of the United Nations (FAO) support this view, reporting that marine fisheries production peaked in the 1980s and that over recent years, approximately half of fisheries have been exploited to their maximum capacity, one quarter overexploited, collapsed or in decline and only one quarter have had potential for increased production.

Inland fisheries have increased throughout the last half century reaching about nine million tonnes in 2002, although this trend has been accompanied in many lake and river systems by overfishing and the collapse of individual large, valuable species. "Ecosystem overfishing" has occurred as the species assemblage is fished down and fisheries use smaller nets to catch smaller and less valuable species. Inland fish stocks have also been aversely affected by pollution, habitat alteration, infrastructure (dams and water management schemes) and introduction of alien species and cultured fish.

In addition to stock collapses, overfishing in general has reduced revenues and economic efficiency, increased variability and reduced the resilience of stocks and catches. The aquatic ecosystems have been profoundly altered by fishing, with a generalised trend of "fishing down the food web" as fish from higher trophic levels decline, leading to lower trophic levels of harvests and a range of ecosystem effects, including disturbance of sensitive habitats by destructive gears such as explosives, poisons and heavy bottom trawling equipment. Extinctions of target fish species, even marine species with high reproductive outputs, are now thought to be possible while impacts on incidentally caught species and habitats also constitute a loss of aquatic biodiversity and can impact ecological processes like predation, bioerosion, provision of food to seabirds and transport of nutrients. By introducing a new and dominant selection pressure, fishing probably also affects the genetic character of fish stocks.

Many industrialized fisheries suffer from over-investment and surplus fishing capacity making it economically and politically difficult to scale back fishing to match biological productivity. Thus, even without any changes attributable to climate change, there is a generally perceived need to reduce fishing capacity and fishing effort in most fisheries.

High profile collapses of Peruvian anchovy stocks, the Northwest Atlantic cod and sea cucumber fisheries throughout the tropical Indian and Pacific oceans are emblematic cases of the failure of fisheries management (in the former cases, in spite of considerable investments in scientific research) and the difficulty of sustainably exploiting many stocks. There is a growing awareness of the importance of understanding human aspects of fisheries and focusing on fisheries governance rather than purely management. Much more attention is now being paid to incentives created by management measures and institutional arrangements around fisheries, including the incorporation of local fishers and their knowledge through co-management and community-based management initiatives. This trend has been accompanied by a greater awareness of the importance of taking account of ecosystems within which fisheries are embedded. Both the involvement of stakeholders and the need to consider the wider ecosystem are incorporated in the Ecosystem Approach to Fisheries.

Another key trend in the nature of fisheries is their increasing commercialization and globalization. Even small-scale fisheries are usually to some extent commercial, involving the sale of at least some of the catch.

Meanwhile, international trade in fisheries products increased sharply until the 1990s. Forty percent of the total value and 33 percent of the total volume of fish produced is traded

internationally. Of this, about half is exported from developing countries earning them greater export revenues than any other food commodity. In the case of specific high value fisheries like sea urchins or live reef fish, demand from markets on the other side of the world can influence fishers in remote areas and result in rapid development, overexploitation and collapse of fisheries within a matter of years.

Marine and freshwater fisheries are susceptible to a wide range of climate change impacts. The ecological systems which support fisheries are already known to be sensitive to climate variability. For example, in 2007, the International Panel on Climate Change (IPCC) highlighted various risks to aquatic systems from climate change, including loss of coastal wetlands, coral bleaching and changes in the distribution and timing of fresh water flows, and acknowledged the uncertain effect of acidification of oceanic waters which is predicted to have profound impacts on marine ecosystems. Meanwhile, the human side of fisheries: fisher folk, fishing communities and related industries are concentrated in coastal or low lying zones which are increasingly at risk from sea level rise, extreme weather events and a wide range of human pressures. While poverty in fishing communities or other forms of marginalization reduces their ability to adapt and respond to change, increasingly globalized fish markets are creating new vulnerabilities to market disruptions which may result from climate change.

A key feature of the socio-economics of inland fisheries, which may influence how they interact with climate change, is the intense seasonality of many highly productive floodplain fisheries, for example those in Southeast Asia (SEA) and Bangladesh. Somewhat related to this trend is the tendency for inland fisheries to be conducted by people who do not define themselves as fishers, but rather engage with seasonal fisheries alongside other livelihood options.

Fisheries Categories

Fisheries demonstrate wide diversity in terms of scale, environment, species, technology, markets, fishers, management arrangements and political contexts and these factors will determine how each is affected by climate change. To simplify this diversity, a generalization will be made between large-scale/industrialized and small-scale/artisanal fisheries. Small-scale fisheries employ more than 99 percent of fishers but produce approximately 50 percent of global seafood catches.

Fisheries for reduction to fishmeal and fish oil are clearly distinguishable from fisheries for food production as they are subject to different market dynamics and have different implications for society.

Inland freshwater fisheries will be distinguished from marine fisheries. Inland fisheries are based on very different biophysical systems to marine fisheries, but in this chapter, which focuses on the impacts of climate change on fisher folk rather than biophysical mechanisms, much of the discussion of vulnerability and poverty will be relevant to small-scale marine fisheries as well as inland fisheries (which are generally small-scale in nature).

Vulnerability and Resilience

Vulnerability has become a key concept in the climate change literature. It is defined as the susceptibility of groups or individuals to harm as a result of climatic changes. Vulnerability is often compounded by other stresses and recognizes that the way in which people and systems are affected by climate change is determined by external environmental threats, internal factors determining the impact of those threats and how systems and individuals dynamically respond to changes. The Intergovernmental Panel on Climate Change definition of vulnerability is "...a function of the character, magnitude, and rate of climatic variation to which a system is exposed, its sensitivity, and its adaptive capacity."

The vulnerability of an individual, community or larger social group depends on its capacity to respond to external stresses that may come from environmental variability or from change imposed by economic or social forces outside the local domain. Vulnerability is complex and depends on a combination of natural and socio-political attributes and geography. Non-climate factors such as poverty, inequality, food insecurity, conflict, disease and globalization can increase vulnerability by affecting the exposure, sensitivity and adaptive capacity of systems, communities and individuals.

Resilience is a concept that is related to vulnerability and adaptive capacity. It has increasingly been applied to the management of linked social-ecological systems (SES) such as fisheries. Resilience is usually applied with an explicit recognition that SES are "complex systems" resulting in uncertain and surprising behaviours including path dependence, alternative stable states, thresholds and periods of apparent stability punctuated by rapid shifts to qualitatively different behaviours. A resilience perspective does not focus on the ability of a system to resist change. Instead it emphasises the importance of disturbance, reorganization and renewal. The dynamic nature of the concept makes it useful when considering uncertain effects of climate change on complex systems like fisheries. Social-ecological resilience includes the importance of social learning, knowledge systems, leadership, social networks and institutions for navigating disturbance, adapting to change and managing the resilience of a system to remain in a desirable state. Accordingly, resilience is seen as the capacity of a system to absorb disturbance while maintaining its basic functions, to self-organise and to build capacity for learning. Resilience of aquatic production in the developing world has been defined as the ability to "absorb shocks and reorganise… following stresses and disturbance while still delivering benefits for poverty reduction."

Socio-economic Context of Fisheries

The poverty of many fishing communities has conventionally been understood as deriving endogenously because of the inevitable overexploitation and poor returns from open-access resources (people are poor because they are fishers); or exogenously because the influx of the poorest of the poor into fisheries as a last resort (they are fishers because they are poor). However, both Bene and Smith, Nguyen Khoa and Lorenzen suggest that this view is over simplistic and small-scale fisheries need to be understood within their wider socio-economic and cultural context. Both authors draw on Allison and Ellis who introduced the analytical framework of the sustainable livelihoods approach to explicitly detail aspects of small-scale fisheries that should be considered.

A livelihood can be defined as the capabilities, assets and activities required for means of living. The concept of sustainable livelihood seeks to bring together the critical factors, assets and activities that affect the vulnerability or strength of household strategies. People can access, build and draw upon five types of capital assets: human, natural, financial, social and physical.

Access to assets is mediated by policies, institutions or processes (PIPs) such as market or organizations. Livelihoods are also affected by a vulnerability context which includes, for instance, seasonality and changes in fuel prices.

This framework and the perspective of fisheries being only one of a variety of sectors which individuals, households or communities draw on for their livelihoods helps to understand some of the linkages of fisheries with wider systems and emphasises the importance of context. This leads to a more holistic analysis of fisheries and climate change because it sees fisheries, not as a simple relationship between a community and an aquatic

production system, but rather as part of a broader socio-economic system which is also affected by climate change. Climate change can be seen to impact each of the five types of assets as well as changing the vulnerability context and impacting on policies, institutions and processes.

Climate Change and Climate Variability

Fisheries have always been affected by variable climate, including rare extreme events such as upwelling failures, hurricanes and flooding. Rather than a steady increase in temperature, climate change is likely to be experienced as an increased frequency of extreme events. Therefore, it is valid to analyse how fisheries react and adapt to existing climate fluctuations. This assumption, that future climate change will be manifested in the form of increasing severity of familiar phenomenon, may be appropriate to guide policy and actions for near-term climate impacts, but it should be borne in mind that thresholds, or "tipping points" may exist, which shift SES into qualitatively different conditions and present novel problems for fisheries sustainability and management.

Units and Scales of Analysis

Impacts of, vulnerability to, and adaptation to climate change can be examined for many different aspects of "fisheries" (e.g. sustainable fish production, well being, economies, food security and livelihoods) at a range of scales (e.g. nations, communities, sectors, fishing operations, households and individuals). Each of these aspects will be affected differently by climate change. For example, stopping fishing as an adaptation to reduced production would be viewed differently from a perspective of sustainable fish production compared to a perspective of the well-being of the communities involved.

The scale of analysis can also affect findings. For example, national-level statistics might identify vulnerabilities of individual economies to certain impacts, but fail to discern vulnerable individuals or social groups within nations that are not highlighted as vulnerable by national statistics. This chapter uses fisher folk and their communities as the main unit of analysis and examines vulnerability at a range of scales.

Fisheries and Climate Change Mitigation

Fisheries' Contribution to Greenhouse Gas Emissions

Fisheries activities contribute to emissions of greenhouse gases (GHG), which are responsible for human-induced climate change, both during capture operations and subsequently during the transport, processing and storage of fish. Most work on fisheries' contribution to climate change has concluded that the minimal contribution of the sector to climate change does not warrant much focus on mitigation, and there is limited information specific to fisheries on contributions to emissions. However, Tyedmers et al. calculate that fishing fleets consume the same quantity of oil as the whole of the Netherlands. This section discusses some of the emission pathways, potential mitigation measures, and examples.

Emissions from Fisheries Operations

Although most fisheries use vessels that are in some ways motorized and powered by fossil fuels, different types of fisheries use different fuels. Small fishing vessels use petrol or occasionally diesel in outboard and inboard engines, while medium-sized fishing vessels use diesel because it is less flammable than petrol. Only the very largest fishing vessels (more than 1 000 tonnes) use the most polluting heavy oil which fuels large freight vessels. This is because the heavy oil requires specialized equipment to treat it before it is passed to the engines

Current estimates suggest that aviation and the world shipping fleet, including commercial fisheries operations, contribute around the same amount of CO_2 emissions. In 2001 the 90 000 or so ships over 100 tonnes in the world fleet, consumed around 280 million tonnes of fuel, with emissions of around 813 Tg CO_2 and 21.4 Tg NO_x (a powerful GHG) in 2000. There were around 23 000 fishing vessels and fish factory ships over 100 tonnes registered in 2001, making up 23 percent of the world's total fleet. Eyring et al., derive emission coefficients for these classes of vehicle, from which we estimate that total emissions from large fishing vessels is around 69.2 Tg CO_2 per annum, representing 8.5 percent of all shipping emissions. This estimate is midway between the higher estimate of Tyedmers, Watson and Pauly, who used FAO catch statistics and typical fuel/catch efficiency for various fisheries to estimate fuel consumption of the global fishing fleet in 2000, and that of FAO which analysed fuel oil use by fishing vessels in 2005.

FAO's estimate is considerably lower, perhaps reflecting reductions in the fishing fleet from 2001 to 2005. However, trends in vessel numbers would not explain the substantially lower estimate because reductions in some areas were compensated for by increases in others. For example, the number and total kW engine power of EU vessels declined by about nine percent, while, in spite of plans to address overcapacity, the size and power of China's fleet increased by seven and nine percent respectively.

In some cases, mobile fishing gears, especially demersal trawls are less fuel efficient than static gears. However, the energy efficiency of individual fishing operations needs to be specifically examined because some industrialized passive gear fisheries can be highly fuel intensive. Fuel costs in 2005 were estimated to be nearly 30 percent of revenue for mobile demersal gears in developed countries. Fleets in the developing world tend to be less fuel efficient in terms of costs and catch revenue, spending up to 50 percent of total catch revenue on fuel. These figures do not allow absolute fuel consumption to be compared because they are affected by variable price of fuel and catch in different fisheries and countries.

Fuel efficiency can be reduced by poor fisheries management. The "race to fish" which can be exacerbated by certain management measures creates incentives to increase engine power. Meanwhile, overfished stocks at lower densities and lower individual sizes require vessels to exert more effort, catch a higher number of individual fish, travel to more distant or deeper fishing grounds and/or fish over a wider area to land the same volume of fish, all of which would increase fuel use per tonne of landings.

Mitigation of Operational Emissions

Increasing fuel costs are likely to continue to pressure the fishing industry to improve fuel efficiency in order to remain profitable. For example, switching to more efficient vessels or gears, such as from single to twin trawls. However, such practices are only estimated to offer a reduction in fuel use of up to 20 percent. Options also exist for small-scale fishers to reduce their fuel use by improving the efficiency of their vessels, using sails or changing fishing behaviour.

Emissions from Trade

FAO estimates that 53 million tonnes of fish were internationally traded in 2004 including products of both fisheries and aquaculture. The transport of this fish will result in emissions of GHGs. High value fish products such as tuna imports to Japan, are frequently transported by air freight and thus would have especially large transport related emissions. Air freight imports of fish to the United States, Europe and Asia are estimated at 200 000, 100 000 and 135 000 tonnes, respectively. Fisheries may make a regionally significant contribution to air freight. For example fish, molluscs and crustaceans were the most frequently airfreighted commodity from New Zealand in 1997, while 10 percent of all air freight from British Columbia in 1996 was fisheries products.

Despite rapid increases in global air freight of fish products until the early 2000s, the quantities seem to have since stagnated. This may be because of competition with other airfreighted commodities, the reluctance of airlines to carry fish and a trend towards transport of fish frozen at source in refrigerated containers. Emissions per kilogram of product transported by air are many times higher than for those transported by sea. Saunders and Hayes estimate coefficients for the transport of agricultural products and the same coefficients should be relevant for fish export (though fish export may be higher if more refrigeration is used). Intercontinental air freight of fish may thus emit 8.5 kg of CO_2 per kilogram of fish shipped, which is about 3.5 times the emissions from sea freight and more than 90 times the emissions from local transportation of fish if they are consumed within 400 km of the source.

Assuming that emissions per kilogram for fish were similar to intercontinental agricultural produce, the 435 000 tonnes of air freighted fish imports to the United States of America, Europe and Asia would give rise to 3.7 Tg CO_2 emissions, which is approximately three to nine percent of the estimates for operational CO_2 emissions from fishing vessels. Emissions from the remaining, non-air freighted 52.5 million tonnes of internationally traded fish depend on the distance and transport mode used.

Clearly, more detailed information on transport modes is needed to provide a reliable estimate of emissions from fish transport, but it is possible that emissions from this sector are as significant as operational emissions. Continuing internationalization of the fish trade will increase fisheries' contributions to CO_2 emissions if transport efficiency and the ratio of air and surface freight remains the same, while increased use of bulk sea-freight or local consumption may reduce the overall emissions from fish transport.

Other Contributions from Fisheries to Mitigation

Some initial research has been conducted into the utilization of waste products from fish processing for producing biodiesel. This may offer alternatives to fossil fuels or terrestrial biodiesels in specific instances where large quantities of fish fats are available. For example, a tilapia processing company in Honduras generates electricity and runs vehicles based on waste fish fat (Tony Piccolo, personal communication). This is based on the utilization of waste products from industrial processing of cultured fish. Given the nutritional value of fish, such uses are unlikely to be desirable in typical capture fisheries unless there are similarly large quantities of otherwise waste fish products.

Impacts of Global Mitigation Actions on Fisheries

Aviation and shipping currently lie outside any emissions trading scheme. Distant water fishing vessels that are supplied with fuel outside territorial waters are therefore not included and can also avoid domestic taxes on fuel. In contrast, vessels fishing within their own country's exclusive economic zone (EEZ) are liable to pay fuel duty and be incorporated into current mechanisms. As the post-Kyoto mechanism for 2012 is negotiated, aviation and shipping may become incorporated with implications for the emissions and fuel use of all fishing vessels.

As the vast majority of fisheries operations are entirely reliant on fossil fuels, they are vulnerable to any decrease in the availability of, or increase in the price of fuel. The doubling of the diesel price during 2004 and 2005, for example, led to a doubling of the proportion of fishers' revenue that they spent on fuel and rendered many individual fishing operations unprofitable.

With 40 percent of fish catch being internationally traded increases in transport and shipping costs (i.e. through carbon taxes or other mitigation measures) will affect markets and potentially reduce the profitability of the sector. This may also affect the food security of poorer fish-importing countries as the costs of importing fish increase.

Impacts by Sector

Climate change can be expected to impact fisheries through a diverse range of pathways and drivers. Effects of climate change can be direct or indirect, resulting from processes in aquatic ecological systems or by political, economic and social systems.

Small-scale and Artisanal Marine Fisheries

The small-scale sector is susceptible to a variety of indirect ecological impacts depending on the ecological system on which the fishery is based. Coral reefs, for example, support small-scale fisheries throughout the tropical western Atlantic, Indian and Pacific oceans and are at risk from elevated water temperatures and acidification in addition to a range of more direct local impacts. The risk of severe bleaching and mortality of corals with rising sea surface temperatures may threaten the productivity of these fisheries. The distribution of coral reefs, coinciding with large numbers of developing country populations in Southeast Asia, East Africa and throughout the Pacific, suggest that many millions of small-scale fishers are dependent on coral reefs for their livelihoods. Nearshore habitats and wetlands, like mangroves and seagrass beds which are often the target areas of small-scale fishers, or which may provide breeding or nursery areas for important species, may be impacted by sea level rise, especially where coastal development restricts landward expansion of the ecosystem.

As species distributions change in response to climate change, small-scale fishers may be less able to adapt by following them because of limited mobility. Traditional area-based access rights institutions will become strained by the loss or relocation of local resources. However, while some fisher folk will see the disappearance of their target species, others could see an increase in landings of species of high commercial value. For example, in the Humboldt Current system during El Niño years, landings of shrimp and octopus increase in northern Peru while in the south, tropical warm-water conditions increase the landings of scallops. These species have higher market values than more traditional species and international markets have developed for them.

Additionally, input of fresh water in estuaries may favour the appearance of brackish water species. For example, during the El Niño of 1997 to 1998, increased rainfall in northern Peru changed salinity patterns in estuaries, favouring the mullet fishery and in Columbia during the La Niña event of 1999 to 2000, a tilapia fishery boom was observed in Columbia. This was caused by salinity changes.

Small-scale fishers are particularly exposed to direct climate change impacts because they tend to live in the most seaward communities and are thus at risk from damage to property and infrastructure from multiple direct impacts such as sea level rise, increasing storm intensity and frequency. Worsening storms also increase the risks associated with working at sea, and changes in weather patterns may disrupt fishing practises that are based on traditional knowledge of local weather and current systems.

Disruption of other sectors (e.g. agriculture, tourism, manufacturing) by extreme events could lead to indirect socio-economic effects. The displacement of labour into fishing can lead to conflicts over labour opportunities and increased fishing pressure. This was observed as a result of hurricanes in the Caribbean. Droughts and resultant agricultural failure forecast in some areas of sub-Saharan Africa may lead to so-called "environmental refugees" moving to coastal areas and creating an influx of surplus fishing labour.

The livelihoods of small-scale fishers are already vulnerable to a range of non-climate risks, including fluctuating resources, loss of access, HIV/AIDS, market fluctuations, conflict, political marginalization and poor governance. This insecurity inhibits investment in long-term strategies for sustainable fisheries and will be exacerbated by additional insecurities caused by climate change impacts. Small-scale fishers also generally lack insurance.

Large-scale Marine Fisheries

Many of the world's largest fisheries (most notably the Peruvian anchoveta – responsible for more than 10 percent of the world's landings) are based on upwelling ecosystems and thus are highly vulnerable to changes in climate and currents. Annual catches of Peruvian anchoveta, for example, have fluctuated between 1.7 and 11.3 million tonnes within the past decade in response to El Niño climate disruptions.

Large-scale changes affect the distributions of species and, hence, production systems. For example, the predicted northern movement of Pacific tuna stocks may disrupt fish-based industries because existing infrastructure (e.g. landing facilities and processing plants) will no longer be conveniently located close to new fishing grounds. In addition, changes in the distribution of stocks and catches may occur across national boundaries.

A lack of well-defined and stable resource boundaries present particular challenges for fisheries governance in the context of climate change. Changes in fish stock distribution and fluctuations in the abundance of conventionally fished and "new" species may disrupt existing allocation arrangements. For instance, changes in Pacific salmon distribution as a result of sea surface temperatures and circulation patterns have led to conflicts over management agreements between the United States and Canada. Similarly, it is forecast that temperature changes in the Pacific Islands could lead to a spatial redistribution of tuna resources to higher latitudes within the Pacific Ocean, leading to conflicts over the stock of tuna between industrial foreign fleets and national ones restricted to their EEZ. Such problems can also occur on subnational scales between local jurisdictions, traditionally managed areas or territorial rights systems.

Rigid spatial management tools, such as permanently closed areas to protect spawning or migration areas, management schemes based on EEZ boundaries or transboundary fisheries management agreements may become inappropriate for new spatial fish stock configurations. Temporal management instruments (e.g. closed seasons) may also become ineffective if the seasonality of target species changes in response to altered climate regimes.

Industrial fisheries are also prone to the direct climate change impacts of sea level rise and increasing frequency and intensity of extreme weather. As with small-scale fisheries, fishing operations may be directly disrupted by poor weather, while extreme events can damage vessels and shore-based infrastructure. City ports and facilities required by larger vessels may be affected. An increasing number of large coastal cities are at risk from sea level rise and extreme weather, especially in rapidly developing Asian economies.

Indirect socio-economic impacts on industrial fisheries may include flooding or health impacts on vulnerable societies which may affect employment, markets or processing facilities. The aquaculture industry is a major market for fishmeal from capture fisheries and climate change impacts may affect markets for reduction fisheries, although current projections are for fishmeal and fish oil demands to continue to increase in the near future.

Positive indirect impacts for some fisheries may result from declines in other fisheries which compete for global markets. For example, while eastern pacific upwelling fisheries were adversely affected in El Niño years, Danish fishers received near record prices for Baltic sprat, a competing species for fishmeal production.

Inland Fisheries

Inland fisheries ecology is profoundly affected by changes in precipitation and run-off which may occur due to climate change. Lake fisheries in southern Africa for example, will likely be heavily impacted by reduced lake levels and catches.

In basins where run-off and discharge rates are expected to increase, the seasonal inundation of river floodplains such as those in the Ganges Basin in South Asia, fish yields may increase as larger areas of ephemeral spawning and feeding areas are exploited by lateral migrant species. In Bangladesh, a 20 to 40 percent increase in flooded areas could raise total annual yields by 60 000 to 130 000 tonnes. However, whilst the discharge rates and flooded areas of many rivers in South and South-East Asia may increase, their dry season flows are often predicted to decline and exploitable biomass is more sensitive to dry, than flood season conditions. Any increases in yield arising from more extensive flooding may therefore be offset by dry season declines. In addition, changes to the hydrological regime and the risk of droughts and flooding may create further incentives to invest in large-scale infrastructure projects like flood defences, hydropower dams and irrigation schemes, which are already known to have complex (and often negative) interactions with fisheries.

The shallow, highly productive Lake Chilwa in Malawi supports a US$10 million a year fish trade. However, rainfall variations have led to periodic drying out of the entire lake and time-series demonstrate that the productivity of the fishery is strongly tied to the amount of water in the lake. During drought periods, some fishers diversified their livelihoods to farming, pastoralism and other occupations, while some wealthier, more specialized fishers, migrated to fisheries in other lakes in the region.

Market and Trade Impacts

Fisheries can be affected by direct climate impacts on processing and trade. For example, following hurricane Katrina, fishers in the Mississippi area of the United States were unable to sell, catch or buy fuel or ice while heavy rain in Peru in 1998 disrupted road networks and prevented rural fishing communities from accessing their usual markets.

Increasing frequency of algal blooms, shellfish poisoning and ciguatera poisoning because of warming seas, ecological shifts and the occurrence of water-borne human pathogens, like Vibrio in areas affected by flooding may lead to fears over fish contamination. These factors may adversely affect fish markets although this impact is still uncertain.

Potential Positive Impacts

In addition to negative impacts, climate change is likely to create opportunities and positive impacts in some fisheries, although these are not well understood or described in the literature. In inland waters, fisheries created by increases in flooded areas may partially offset the loss of land for agriculture or other economic activities. In Peru, increased sea surface temperatures negatively affect pelagic fisheries for small-scale artisanal fishers, but also bring a variety of (sub) tropical immigrants and expands the distribution zone of some species, illustrating very well how climate change could bring new opportunities to fisher folk and their communities. Indeed, during the El Niño of 1982 to 1983 and 1997 to 1998, penaeid shrimps and rock lobsters from the Panamic Province appeared in Peru. These species, along with dolphin fish (mahi-mahi), tuna and diamond shark created a new economic opportunity for the artisanal fishing sector.

An extreme case is the potential creation of an entirely novel open water fishery as a result of the melting of the Arctic Ocean. The management of as yet nonexistent fisheries with no prior governance arrangements provides a challenge in terms of uncertainty and lack of experience, but also an opportunity to develop governance and management with precautionary limits before overcapacity develops.

The adaptive capacity of economies, fishing sectors, communities, individuals and governance systems will determine the extent to which they are able to maximize the opportunities created by new fisheries.

Observed and Future Impacts

Observed Impacts of Climate Change and Variability

Many fisheries are known to be profoundly controlled by climate variability through ecological impacts. Meanwhile, long-term climate-related changes have been observed in marine ecosystems including in targeted fish populations. However, in spite of the ecological changes that have been recorded, impacts on fisheries have largely yet to be discerned from pre-existing variability and non-climate impacts (of overexploitation, market fluctuations etc.). Even fisheries associated with coral reefs that have been profoundly impacted by climate change have yet to demonstrate a significant impact. Although a lowering of ocean pH of 0.1 unit has been observed since 1750, no significant impacts of acidification on fisheries have yet been observed although long-term forecasts are alarming.

Coastal zones throughout the world are experiencing erosion, threatening coastal communities with flooding and loss of coastal ecosystems. A variety of processes are responsible for this, including changes in land use. However, erosion may also be exacerbated by climate-mediated sea level rise, although the complexity of coastal dynamics makes it difficult to isolate the impact of climate change.

Likely Additional Impacts Within the Next 50 Years

Existing climate trends will increase over the next century and are expected to impact more severely on aquatic ecosystems and, directly and indirectly, on fishing sectors, markets and communities. Loss of corals through bleaching is very likely to occur over the next 50 years, with consequent impacts on the productivity of reef fisheries and potentially on coastal protection as reefs degrade. Sea level will continue to rise and by 2100 will have increased by a further 20 to 60 cm, leading to elevated extreme high sea levels, greater flooding risk and increased loss of coastal habitats.

In addition to incremental changes of existing trends, complex social and ecological systems such as coastal zones and fisheries, may exhibit sudden qualitative shifts in behaviour when forcing variables past certain thresholds. In addition to this non-linearity in systems, assumptions of gradual change may be based on an incomplete understanding of the mechanisms which will lead to more rapid shifts. For example, IPCC originally estimated that the Greenland ice sheet would take more than 1

000 years to melt, but recent observations suggest that the process is already happening faster owing to mechanisms for ice collapse that were not incorporated into the projections. Similarly, predictions of changes to fisheries' social and ecological systems may be based on inadequate knowledge of mechanisms and potential "tipping elements", which might be responsible for sudden or irreversible changes. Climate change may, therefore, result in sudden, surprising and irreversible changes in coastal systems. The infamous collapse of the Northwest Atlantic northern cod fishery provides a (non-climate-related) example where chronic overfishing led to a sudden, unexpected and irreversible loss in production from this fishery. Thus, existing observations of linear trends cannot be used to reliably predict impacts within the next 50 years.

Impacts of Climate Change in the Context of Other Trends

Future impacts of climate change on fisheries need to be seen in light of the considerable changes which might be expected within society regardless of climate change, for example in markets, technology and governance. This evolving context for fisheries may mean that the impacts of climate change cannot be predicted by analysing how fisheries systems in their contemporary state will be affected by future climate change. It is likely that in the future, climate change will impact on future fisheries in different configurations from the current situation. For example, if fisheries are better managed in the future through incentive-based and participatory management of the ecosystem and with more efficient enforcement, then fish stocks will be better able to withstand biophysical impacts on recruitment and fisheries ecosystems will be more resilient to changes. In a world in which demand for fish incrcascs, priccs continuc to risc and fisheries become increasingly globalized, commercial fisheries may be able to maintain profitability in the light of declining yields. However, subsistence fisheries and local markets in poorer countries may become more sensitive to economic demand from richer countries and as more fish production is directed to exports, the contribution of fisheries to food security may decline in poorer countries.

Synergistic Impacts

Literature on climate change impacts necessarily tend to list separate impacts but it is important to be aware of potential synergistic and cumulative effects of multiple impacts.

Uncertainty of Impacts

Climate change is occurring and an increasing range of observed impacts, there is still considerable uncertainty in the extent, magnitude, rate and direction of changes and impacts. Meanwhile, unlike for terrestrial systems supporting agriculture, there is a lack of quantitative predictions of climate effects on aquatic systems. The relative importance of different impacts and potential interactions between them are very poorly understood and the uncertainty in predictions about climate variables is amplified by poorly understood responses of biophysical systems. A further complexity and unpredictability is in how people and economies, and their complex relationships with local ecosystems might respond to change. This underscores the need for social scientists as well as economists and natural scientists to be engaged in policy recommendations and management. It also emphasises the need for fisheries governance regimes to be flexible enough to adapt to and learn from unforeseen changes (i.e. to have high adaptive capacity). Frameworks such as adaptive co-management are being developed and may provide some of this flexibility but as yet they have not been fully tested on a larger scale.

Vulnerability of Regions, Groups and Hot Spots

Climate change impacts on fisheries will have uneven effects on different geographic areas, countries, social groupings and individuals. Vulnerability depends not only on the distribution of climate impacts (exposure) but on their sensitivity and adaptive capacity. Thus vulnerability is socially differentiated: virtually all weather-related hazards associated with climate variability, as well as human causes of vulnerability, impact differently on different groups in society. Many comparative studies have noted that the poor and marginalized have historically been most at risk from natural hazards and that this vulnerability will be amplified by climatic changes. Poorer households are, for example, forced to live in higher risk areas, exposing them to the impacts of coastal flooding and have less capacity to cope with reduced yields in subsistence fisheries. Women are differentially at risk from many elements of weather-related hazards, including, for example, the burden of work in recovery of home and livelihood after a catastrophic event.

Assessing the vulnerability of different geographic areas, countries, social groupings and individuals, aims to identify those who will be most adversely affected, which information can be used to guide policy and interventions to assist adaptation.

Geographic Regions with High Potential Exposure

The greatest warming of air temperatures thus far has been experienced in high latitudes and this is likely to continue with future climate change. However, changes in water temperatures are less well predicted and are mediated by ocean currents. Only some climate impacts on fisheries are mediated by temperature, so projected air temperature changes commonly presented in climate forecasts are a poor measure of potential exposure. Low latitude regions, for example, where fisheries rely on upwellings, coral reef systems or susceptible fresh water flows may be more exposed to climate impacts than high latitude regions where most warming is predicted.

IPCC predictions suggest that tropical storm intensity will increase, specifically impacting fishing communities and infrastructure in tropical storm areas. It is also possible, but less certain, that the existing tropical storm belt will expand to affect more areas. In this case, communities for whom tropical storms are a novel disturbance may initially be more sensitive if they lack appropriate infrastructure design, early warning systems and knowledge based on previous experience.

Fisheries communities located in deltas or on coral atolls and ice dominated coasts will be particularly vulnerable to sea level rise and associated risks of flooding, saline intrusion and coastal erosion.

Vulnerable Economies

Developing countries in tropical regions are usually assumed to have lower adaptive capacities than countries with high levels of economic and human development. This is because of lower availability of resources and institutions necessary to facilitate adaptation.

A national level analysis of the vulnerability of 132 economies to climate impacts on fisheries used predicted climate change, the sensitivity of each economy to disruption to fisheries and adaptive capacity, as indicated by statistics on development and GDP. According to the resultant index, countries in western and central Africa (because of low levels of development and high consumption of fish), northwest South America (due to very large landings) and four Asian countries were most vulnerable. Russia and Ukraine were the only two high latitude countries identified with high vulnerability due to the high degree of expected warming and low adaptive capacity scores.

The analysis highlighted the importance of low adaptive capacity for elevating the vulnerability of African countries even though greater warming is predicted at higher latitudes. While the analysis was a pioneering study of vulnerability of fisheries to climate change, there are several limitations. Firstly, projected increase in air temperature was assumed to be an indicator of exposure to climate change, whereas extreme events or non temperature mediated impacts may be most important. Secondly, data availability prevented the inclusion of most small island developing states, expected to be vulnerable because of a high reliance on fisheries, low adaptive capacity and high exposure to extreme events. Finally, analysis at the national scale required crude generalizations about countries which may miss sub national hotspots of vulnerable sectors or communities.

To improve large-scale mapping of vulnerability, more detailed predictions of changes in the likelihood of extreme events, hydrology and oceanography are needed to better characterise exposure. Integrative earth science and ecology projects such as the United Kingdom Natural Environmental Research Council (NERC)'s Quest-Fish project will make some advances to better characterizing aspects of exposure to move beyond use of projected air temperature changes. Meanwhile, higher resolution, sub national data on resource use, fish consumption and trade, fisheries production and poverty will allow more detailed mapping of sensitivity and adaptive capacity.

Vulnerability of Communities

Vulnerability can also be analysed based on statistics at the sub national level. For example, McClanahan et al. derived an index of adaptive capacity with respect to a loss of fishing livelihoods of 29 coastal communities in five nations in the western Indian Ocean. The index combined eight variables proposed to be important for adaptive capacity weighted according to relative importance as judged by experts from across the region. The resultant ranking of communities could broadly have been predicted from national level development statistics but exceptions include communities in Madagascar (with the lowest development status of the five nations), which score more highly than communities in richer countries because of high occupational mobility, "decline response" and "social capital" (e.g. Sahasoa).

Thus a range of factors indicate adaptive capacity, and wealth may not be a complete indicator.

Vulnerable Groups within Society

At even finer scales, vulnerability varies between individuals within a community, with some groups particularly vulnerable.

Vulnerability is often assumed to be generally correlated with poverty. Hurricane Katrina, which hit New Orleans in August 2005, demonstrated how the poor are particularly vulnerable, even in the most prosperous countries. Poor families, including a high proportion of African Americans, were less likely to evacuate in advance of the hurricane leading to higher death tolls and subsequent impacts on housing, education and psychological state. Poorer members of communities are also least likely to have insurance or access to early warning information.

In addition to vulnerability to disasters, the poorest members of society are generally assumed to have less adaptive capacity to cope with gradual changes or declines in livelihoods. For example, in a fisheries context, Kenyan fishers from poorer households were more likely to be trapped in a declining fishery.

Individual factors other than poverty can also affect vulnerability. For example, women are more vulnerable to natural hazards and climate change impacts owing to increased likelihood of being around the home and increased burdens of care after hazards. It is also assumed for many societies that women possess lower levels of adaptive capacity to men. For example, they have fewer economic options, generally lower education attainment, a greater lack of rights and access to resources and may be more likely to endure the burden of care after hazards. Women headed households, which tend to be among the poorest households in many societies are considered especially vulnerable. The importance of contextual factors is well illustrated by studies of the impacts of the 2004 Indian Ocean Tsunami. Throughout the affected coastal region, many more women than men were killed; in some communities two to three times more women than men. A range of factors made women more vulnerable and some were very locale and context specific. For example, they included ability to swim, physical strength and the need to protect and care for children and elderly. At some locations, because of women's roles in processing and marketing fish, women were waiting on the shore for fishing boats to return at the time of day the tsunami struck. Because of this they suffered higher levels of mortality than the men at sea. Of course these differential deaths have significant implications for relief and rehabilitation and long-term impacts on families and communities. However, there are relatively few rigorous empirical studies, so the literature abounds with generalizations and unproven assumptions. In many situations, for example, women may have access to abundant and diverse forms of social capital which may provide excellent support to overcome certain types of impacts or extreme events.

Gaps in Knowledge about Vulnerability

The ability to identify those most vulnerable to climate change is limited by the lack of high resolution data at appropriate scales and by uncertainty as to the processes that make people and places vulnerable. The IPCC Fourth Assessment highlighted that, in terms of impacts and adaptation, knowledge, monitoring and modelling of observed and future impacts is skewed towards developed nations.

Changing resource scarcity or unpredictability as a result of climate change will clearly affect those whose entire livelihoods are directly dependent on fisheries. But it is unclear whether such dependence on fisheries will underpin efforts to attain sustainable management (as observed in some circumstances and explained by commons management theory); will result in greater overexploitation as future availability becomes uncertain; or will lead to an emphasis on diversification out of fisheries based livelihoods altogether, which may have significant social and even environmental impacts. All three generic responses are likely to occur. Hence defining the goals of desirable and sustainable adaptation for different stakeholders is an important research task for regions at risk.

There is a lack of understanding of how adaptation strategies in general, in coastal areas affected by multiple impacts of climate change, may impact other strategies and neighbouring coastal areas. For example, it has been shown that flood mitigation measures in Bangladesh to protect farmland may negatively affect fisheries. Similarly, hard engineering coastal protection can impact on sediment loading and coastal dynamics in neighbouring coastal areas or countries. And increased "roving" of commercial fishing fleets as stocks migrate will have impacts on neighbouring or even distant countries.

Finally, there may be major thresholds in ecological and physical systems in oceans and coastal areas that directly affect vulnerability of these regions. These include stock collapse thresholds, ocean acidification and its impact of calcifying organisms and rises in temperature above a threshold for mass coral bleaching. The risk of such major shifts in ecology increases the exposure and vulnerability of dependent communities, but may not be known until after a threshold is passed.

Adaptation of Fisheries to Climate Change

Adaptation to climate change is defined in the climate change literature as an adjustment in ecological, social or economic systems, in response to observed or expected changes in climatic stimuli and their effects and impacts in order to alleviate adverse impacts of change, or take advantage of new opportunities. In other words, adaptation is an active set of strategies and actions taken by people in reaction to, or in anticipation of, change in order to enhance or maintain their well-being. Adaptation can therefore involve both building adaptive capacity to increase the ability of individuals, groups or organizations to predict and adapt to changes, as well as implementing adaptation decisions, i.e. transforming that capacity into action. Both dimensions of adaptation can be implemented in preparation for, or in response to impacts generated by a changing climate. Hence adaptation is a continuous stream of activities, actions, decisions and attitudes that informs decisions about all aspects of life and that reflects existing social norms and processes. There are many classifications of adaptation options summarised in Smit et al. based on their purpose, mode of implementation, or on the institutional form they take.

Coulthard highlights the difference between adaptations in the face of resource fluctuations that involve diversifying livelihoods in order to maintain a fishery-based livelihood, and those which involve "hanging up our nets", exiting fisheries for a different livelihood source. Another response often observed during the development of a fishery to cope with reduced yield is to intensify fishing by investing more resources into the fishery. This can be in terms of increasing fishing effort (by spending more time at sea), increasing fishing capacity (by increasing the number, size or efficiency of gears or technology) or fishing farther or deeper than previously. Such adaptation responses obviously have potentially negative long-term consequences if overexploitation is a concern in the fishery. The state of many of the world's fisheries offers little opportunity for sustainable intensification of fishing as an adaptation strategy.

Inevitably adaptation strategies are location and context specific. Indeed, Morton argues that both impacts of and adaptation to climate change, will be difficult to model and hence predict, for smallholder or subsistence agricultural systems. This is because of factors such as the integration of agricultural and non agricultural livelihood strategies and exposure to various stressors, ranging from natural stressors to those related to policy change. The same conditions are likely to prevail in the subsistence fisheries sector, though this has not been researched in the same manner as marginal and subsistence agricultural systems. Faced with this complexity there have been various suggestions and typologies of how adaptation actually occurs for such livelihoods.

Adaptation responses can be conceptually organized based on timing and responsibility. Specific adaptations of industrialised fisheries are likely to differ from those of small-scale fisheries. For example Thornton et al. suggest that intensification, diversification and increasing off farm activities are the most common adaptations in pastoralist settings, while Eriksen et al. observe, in addition, the use of greater biodiversity within cropping systems and use of wild foods. In fisheries, analogous responses can be seen as intensifying fisheries, diversifying species targeted or exiting fishing for other livelihoods. Agrawal and Perrin examine strategies for subsistence resource dependent livelihood systems and suggest all involve functions that pool and share risks through mobility, storage, diversification, communal pooling and exchange. Although most fisheries (even small-scale) are not purely subsistence, this typology of adaptation may be useful for conceptualising small scale fishery adaptations to climate change.

Examples of Adaptation in Fisheries

Fisher folk and their communities around the world are already constantly adapting to various forms of change. Thus, much can be learned by examining how fishers have adapted to climate variability such as El Niño and non climate pressures and shocks such as lost markets or new regulations.

Responses to direct impacts of extreme events on fisheries infrastructure and communities are believed to be more effective if they are anticipatory as part of long-term integrated coastal and disaster risk management planning. Adaptations to sea level rise and increased storm and surge damage include hard (e.g. sea walls) and soft (e.g. wetland rehabilitation or managed retreat) defences, as well as improved information systems to integrate knowledge from different coastal sectors and predict and plan for appropriate strategies.

Indirect socio-economic impacts are arguably less predictable, making it more difficult to discuss specific adaptation measures. Diversified products and markets would make fisheries less prone to economic shocks, while information technologies are becoming more available to small-scale fishers and may help them to navigate international markets and achieve fair prices for their fish. Generally decreasing the marginalization and vulnerability of small-scale fishers is thought to be an anticipatory adaptation to a range of threats, as well as facilitating sustainable management.

Cultural and socio-economic aspects limit people's adaptive capacity in apparently unpredictable ways. In Pulicat Lake in India, for example, access to fish and prawn fisheries is mediated by caste identities. The non fishing caste members do not have traditional hereditary rights of access and subsequently tend to be economically poorer and more marginalized. However, in the face of declines in catches, these non fishing caste fishers were more adaptable to do jobs outside of the fisheries sector. Hence, they had a greater adaptive capacity and were in many ways less vulnerable to annual fluctuations in stocks.

Adaptation of Fisheries Management

Much fisheries management is still loosely based on maximum sustainable yields or similar fixed ideas of the potential productivity of a stock. For example, North Sea groundfish fisheries have recently been managed in order to recover cod to a target biomass of 150 000 tonnes. Although climatic influences on cod productivity are recognised, there is currently no formal strategy by which environmental processes can be incorporated into management targets and measures. As climatic change increases environmental variation, more fisheries managers will have to explicitly consider such variations and move beyond static management parameters for particular stocks. Such changes create an additional imperative to implement the ecosystem approach to fisheries (EAF), a holistic, integrated, and participatory approach to obtain sustainable fisheries.

The role of institutions in adaptation

Institutions, in the broadest sense, mean formal and informal traditions, rules, governance systems, habits, norms and cultures. A technical approach to adaptation can underestimate the importance of institutions (especially informal) to facilitate or limit adaptation. For example, traditional practises or links with alternative livelihoods can be drawn on to adapt to declining fish yields, while cultural identities connected with fishing may limit adaptation, in terms of leaving fisheries, that fisher folk are willing to consider. An extensive literature documents examples of local resource management institutions that facilitate management of common pool resources and it is proposed that such institutions allow adaptive and sustainable management. However, in the face of increasing climate change impacts they can also be a barrier to the flexibility needed for adaptive management. Formal institutions can also constrain adaptation, for example in Peru, the establishment of access rights institutions to improve management of scallop stocks may prevent future migration responses to El Niño shocks), while increasing regulation of gears and sectors in Newfoundland fisheries meant that when cod stocks collapsed, cod fishers who previously exploited a range of species, were "locked-in" to the collapsed cod fishery and unable to benefit from expanding shellfish fisheries.

Building Adaptive Capacity in Fisheries

Uncertainty, surprise and the need for general adaptive capacity

There is great uncertainty in the nature and direction of changes and shocks to fisheries as a result of climate change. Investments in generic adaptive capacity and resilient fisheries systems seem to be a good strategy to support future adaptations which are not currently foreseen. Better managed fisheries with flexible, equitable institutions are expected to have greater adaptive capacity. For example, implementation of the EAF could make an important contribution to adaptation in preparation for the effects of climate change. Many fishers are vulnerable to a range of disturbances which together decrease their adaptive capacity in the face of climate change impacts. Thus, for example, working to address the marginalization of fishing communities and their vulnerability to HIV/AIDS and other diseases and resource insecurity can be seen as a form anticipatory adaptation to climate change shocks.

Have we been here before?

Good management for sustainable stocks, enhanced wellbeing and reduced vulnerability of fisher folk will increase generic adaptive capacity. Therefore, working towards equitable and sustainable fisheries, which has been a goal of fisheries management, may be seen as advancing the adaptive capacity of fishing communities. It has also long been recognized that fisheries management must take account of inherent uncertainty within fisheries which results from climate variability, variable recruitment and unknown linkages within the ecological and social aspects of fisheries.

Thus, adaptation for climate change, in terms of building the resilience of fish stocks and communities and taking account of uncertainty, could be seen as implementation of good fisheries governance as recommended over the past decade, irrespective of climate change, which raises the question of whether new interventions are required to assist adaptation.

Despite the familiarity of the challenges, increased resources and efforts are likely to be needed to adapt fisheries in the face of climate change. The majority of fisheries are still not managed in a sustainable, equitable fashion that takes due account of uncertainty; sudden shifts in systems may result from climate change presenting new challenges; and the magnitude of change may simply overwhelm current options for "good fisheries governance". There may be a need for focused adaptation for poorer, marginalized and most vulnerable fisher folk and communities, which would go beyond previous international development assistance. International financing mechanisms exist and are being developed to support adaptation under the United Nations Framework Convention on Climate Change (UNFCC). These have, for example, funded the creation of National Adaptation Programmes of Action (NAPAs) in poor countries. Significant funds are therefore becoming available for targeted adaptation, but these are thought to be inadequate to address the massive costs of adaptation, while issues of defining and funding adaptation to climate change as distinct from general building of adaptive capacity complicate the process of allocating funds for adaptation.

The priority responsibility for governments, civil society and international organizations with regard to climate change, is to aggressively pursue reductions in greenhouse gas emissions (GHG), because the long-term consequences of climate change are highly complex, unknowable and potentially irreversible and many already marginalised groups appear most vulnerable to its impacts. Fisheries make a moderate contribution to GHG emissions through fossil-fuel-based catching operations and transportation, which may be reduced with improved technology and management of stocks. Previous global emissions already mean that climate change will affect marine and freshwater systems and fishing communities. Governments therefore have a responsibility to facilitate adaptation, especially for groups vulnerable because of their exposure, sensitivity or lack of adaptive capacity. A research imperative is therefore to:

- identify the most vulnerable individuals and communities;
- investigate possible government facilitated adaptation;
- consider constraints on private adaptations; and
- seek desirable adaptations which contribute to long term reductions in vulnerabilities, rather than short-term coping strategies which may enhance vulnerability.

Reviewing the potential impacts of climate change on fisheries suggests a role for public policy in adaptation: to reduce vulnerability, to provide information for planning and stimulating adaptation and to ensure that adaptation actions do not negatively affect other ecosystem services and the viability of fisheries in the long run.

The first rationale for promoting adaptation is to protect those parts of the fishing sector and communities in coastal areas that have the least ability to cope. Coastal regions facing climate change for example are subject to multiple stresses associated with globalization of fisheries, and in the case of developing countries, lack of public infrastructure, high disease burden and many other factors that limit the ability to adapt.

The second public policy response is the provision of high quality information on the risks, vulnerability and threats posed by climate change. Such information includes scenarios of change at the global scale, but it also involves significant investment in incorporation of climate information into coastal land use planning and other forms of regulation. Hence the need for policy integration across government sectors, such as coastal planning, river basin management, agriculture, fisheries themselves and health and nutrition where climate change risks interact.

The third area of public policy response is in the provision and enhancement of the public good aspects of fisheries and related biodiversity and ecosystem services. The Millennium Ecosystem Assessment highlighted the importance of ecosystem services for human wellbeing. Climate change impacts represent enhanced reasons for sustainable fisheries management and incentives to promote biodiversity conservation within coastal regions, given the potential for habitat decline and species extinction throughout the world.

There is already an imperative to improve fisheries governance to take account of natural variability, uncertainty and sustainability and to address overcapacity and overfishing, which lead to economic losses, endanger future fisheries and degrade aquatic ecosystems.

In addition, pro-poor governance of small-scale fisheries is now promoted by international organizations to address marginalization of fishers and equity. These familiar challenges for governance will continue and perhaps become more imperative in the face of climate change. Variability and uncertainty, which have historically been important factors that managers have struggled to take account of, will become more prevalent under climate change. Meanwhile poverty in small-scale fisheries and marginalization of fishers reduces their adaptive capacity.

The wider context of fisheries is also important because of the ways in which politics, socio-economics, demographics, ecology and markets can influence fisheries but also because they are evolving rapidly with processes of globalization. Future climate change will not interact with fisheries in the way it would today because it will affect future fisheries within a future context. This creates additional uncertainty and emphasizes the need for adaptive governance as well as integration of fisheries with other linked sectors, particularly agriculture, which may itself affect fisheries due to climate impacts and adaptation.

Current problems with fisheries management call for strong and reliable institutions governing rcsourcc usc but, paradoxically, top down or rigid approaches which may seem attractive may not offer the flexibility to ensure resilient fisheries systems and communities under climate change. Approaches such as adaptive co-management, proposed to address uncertainty and harness the knowledge and commitment of resource users at multiple scales may offer the best hopes for resilient fisheries. Experiments with such approaches should be extensively trialled and analysed as a priority. Governance systems with a focus on continual learning from experience, which openly treat policy as experimentation, will be more likely to address new challenges as they arise. Policies which place too much emphasis on stability, certainty and top down control may lead to unexpected consequences and may "lock in" fisheries, preventing desirable and sustainable adaptation.

The process of fisheries and their associated communities adapting to climate change is facilitated and constrained by various social factors and involves value-based decisions and trade-offs. Abandoning fisheries as a livelihood may become a necessary reality in some fisheries. The political and value laden nature of adaptation emphasizes the need for equitable and just deliberative processes, for example, if there is a trade-off between actions and policies that assist the most vulnerable and those which provide optimally efficient adaptation or large-scale resilience.

References

Garcia, S.M. & Grainger, R.J.R. 2005. Gloom and doom? The future of marine capture fisheries. *Philosophical transactions of the Royal Society of London – Series B: biological sciences.*

Jentoft, S. 2006. Beyond fisheries management: the phronetic dimension. *Marine Policy,* 30(6): 671- 680.

Miller, K.A. 2007. Climate variability and tropical tuna: management challenges for highly migratory fish stocks. *Marine Policy.*

Morton, J.F. 2007. The impact of climate change on smallholder and subsistence agriculture. *Proceedings of the National Academy of Sciences.*

6

Ecosystem Approach to Capture Fisheries

Aquatic ecosystems, including rivers, lakes and inland seas, flood plains, coastal lagoons and estuaries, coastal shelves and open oceans cover a very large part of the earth's surface and, among other amenities, goods and services, sustain the production of fisheries and aquaculture. They yield about 120 million tonnes of fish and fishery products per year - the largest source of wild protein - and provide a livelihood to as many as 140 million people. Fisheries and aquaculture exploit a large diversity or organisms ranging from algae, ascidians and sea-cucumbers to molluscs, crustaceans, fish and marine mammals.

Most aquatic ecosystems are unavoidably affected by fishery activities that involve a selective removal of part of the natural productivity for human subsistence, economic returns and development. However, undesirable fishing practices in some cases, such as overfishing and use of destructive methods, are unduly affecting these precious ecosystems, calling for urgent corrective action.

Except in the high seas, these ecosystems are also usually used for other purposes such as conservation (e.g. wetlands), forestry (e.g. mangroves), agriculture (e.g. floodplains), offshore mining and oil and gas extraction, and human settlements (e.g. coastal areas). Unfortunately, aquatic ecosystems are also, most often, the ultimate recipients of the pollution produced by human settlements and industrial activities, inland, on the coastal area as well as at sea. Even the most remote areas (e.g. deep ocean and polar seas) are now affected, seriously putting in question the sustainability of present practices and the present ecosystems resources to future generations

Impacts from Fisheries on the Environment

Impacts from fisheries on the environment have been abundantly described and reviewed. More specifically, capture fisheries impact target resources. They reduce their abundance, spawning potential and, possibly, population parameters (growth, maturation, etc.). They modify age and size structure, sex ratio, genetics and species composition of the target resources, as well as of their associated and dependent species. When poorly controlled, fisheries develop excessive fishing capacity, leading to overfishing, with major ecosystem, social and economic consequences.

Fishing may also affect ecological processes at very large scale. The overall impact has been described as comparable, in aquatic systems, to that of agriculture on land in terms of the proportion of the system's primary productivity harvested by humans. Overfishing transforms an originally stable, mature and efficient ecosystem into one that is immature and stressed. This happens in various ways. By targeting and reducing the abundance of high-value predators, fisheries deeply modify the trophic chain and the flows of biomass (and energy) across the ecosystem. They can also alter habitats, most notably by destroying and disturbing bottom topography and the associated habitats (e.g. seagrass and algal beds, coral reefs) and benthic communities.

The alteration of the habitat by various human activities may be physical (e.g. by adding artificial structures like artificial reefs, oil rigs, aquaculture installations), mechanical (e.g. through the "ploughing" effect of dredges and trawls), or chemical (e.g. through injection of nutrients, pesticides, heavy metals, drugs, hormones). Fishing may result in changes in productivity of resources (some positive and some negative) and affects associated species. Some aspects of fisheries can have significant and long-lasting effects:

- destructive fishing techniques using dynamite or cyanides or inadequate fishing practices (e.g. trawling in the wrong habitat); pollution from fish processing plants;
- use of ozone-depleting refrigerants;
- dumping at sea of plastic debris that can entangle marine animals or be swallowed by turtles;

— loss of fishing gear, possibly leading to ghost fishing;

— lack of selectivity, affecting associated and dependent species, resulting in wasteful discarding practices, juvenile mortality, added threat to endangered species, etc.

Poorly-managed large-scale mariculture can damage coastal wetlands and nearshore ecosystems, often used as nurseries by key capture fishery resources, and contribute to ecosystem contamination with food residues, waste, antibiotics, hormones, diseases and alien species.

The Law of the Sea provides that fisheries management must take care also of associated and dependent species. The impact of fishing on these species has been documented in some areas but is still frequently unknown or only partly understood. The decline of primary productivity consumers low in the food chain removes important forage species needed higher in the food web, with cascading effects for the ecosystem. Conversely, the removal of top predators such as mammals, tuna or sharks, may release an unusually large abundance of preys at lower levels with cascading and feedback effects on the food chain and species composition. For example, as most sharks and some batoid fishes (angel fishes) are predators located at or near the top of marine food webs, their depletion modifies the intricate trophic interactions of their ecosystems. The removal of predators through fishing in Kenyan reefs resulted in the expansion of sea urchin population, which apparently led to a decrease in live coral and to loss of topographic complexity, species diversity and fish biomass.

Goñi reports that the hunting of sea otters (Enhydra lutris) in the Northeast Pacific caused a large-scale expansion of sea urchins, the increased grazing of which caused the decline of the important kelp forest. She also reports that, in the Bering Sea, the expansion of the fisheries on pollock (Theragra chalcogramma) during the 1970s has been considered as a probable cause of the decline of several populations of marine mammals, e.g. sea lions (Eumetopias jubatus) by 76%, seals (Callorhinus ursinus) by 60% and (Phoca vitulina) by 85%, as well as the decline of several seabird populations (Urea algae, U. lomvia, Rissa brevirostris, R. tridactyla). All these non-fish species compete directly with the Pollock fisheries since the target species represent 21-90% of their diet.

Impacts can be particularly serious on cartilaginous fish populations, which have a lower productivity and resilience than bony fishes. As a consequence, fisheries targeting shark have a low record of sustainability and some species of skates, sawfish and deep-water dogfish have been virtually extirpated from large regions. A well documented example of direct impact on benthic species is that of modern towed gear (trawls and dredges) which caused, inter alia, long-term changes in abundance and species composition in the Wadden Sea and Australia.

The mortality of benthic species associated with or preyed upon by target bottom fish resources resulting from the use of trawls can vary greatly, depending on how the gear is built or rigged. For example, the addition of a tickler chain on a commercial beam trawl will allow it to catch more of its bottom-fish target but, at the same time, will detach and uproot more benthic species. Likewise, some gear modification can reduce fishing mortality. For example, operating a semi-pelagic trawl 15 cm above the sea bottom has no measurable effect on the benthos community, while the standard demersal trawling dragging on the bottom reduces benthos density by 15.5%. Finally, fishing can have significant impact on the genetic diversity of resources and can permanently change populations characteristics.

Fishing gear can change the living and non-living environment within which the target and other related resources live. Environmental damage may come from the very nature of the fishing technology (e.g. in the use of dynamite or poison) or from the inappropriate use of an otherwise acceptable gear (e.g. using trawls in coral reefs or seagrass beds). The use of dynamite and other explosives for "blast fishing" is still common in parts of Asia, Africa, Caribbean and South Pacific. A relatively small explosive is capable of destroying a three-metre circular area of stony corals. These practices are generally officially banned by fisheries regulations and laws but often persist because the people involved have little, if any, alternative livelihood.The impact on the habitat depends on the gear and sediment type. Highly dynamic, soft bottoms (e.g. coarse sand, hydraulic dunes) may suffer limited damage even when exploited by heavy (including hydraulic) dredges. On the contrary, stable, hard, and highly structured habitats (such as coral reefs, seagrass beds, sponge beds) will be easily damaged. One well-documented example is the use of modern towed gear (trawls and dredges) which caused, inter alia, destruction of seagrass beds (Posidonia oceanica) in the Mediterranean and destruction of the oyster (Cassostrea virginica) habitat in Chesapeake Bay.

Damage is also related to fishing frequency, gear weight and rigging. Addition of heavy tickler chains to the trawl ground rope increases bottom abrasion and turbidity while adding rollers reduces it. The use of sodium cyanide in the Philippines to catch marine tropical fishes for the aquarium trade has led to the destruction of the coral reef habitat and decline of aquarium and food fish.

Impacts of Poor Selectivity, Bycatch and Discards

Fishing generates bycatch and discards. A first attempt to address the issue at global level was made in the late 1990s by FAO and the first estimate of the global extent of the problem (about 27 million tonnes of resources dumped per year) was published by this Organisation. A recent review of the issue has been undertaken by Cook. Most fishing activities are not selective enough to remove from the ocean only the desired targets and will probably never be. This leads to accidentally catching other species (bycatch), part of which has little or no use (at least in the local context) and will be dumped overboard (as discards) together with the offal from fish processing at sea.

The effect is to increase availability of food to scavenger species (including sea birds) and, when concentrated over time, may cause local anoxia of the seabed environment. The resulting amount of organic material may not be negligible. On the North Pacific shelf and upper slope, the amount of offal generated by at-sea production of surimi (a protein extract of fish) is very significant, since the technology used extracts less than 50% of the wet weight from the total catch, the rest being dumped.

In the North Sea, 6.5 to 12.5% of the groundfish caught is dumped at sea. Some of this is consumed by sea birds, but a certain amount of offal becomes available to benthic scavengers. The increase in abundance of dogfish (Scyliorhinus canicula) in northern Spain fisheries and that of Raja radiata in Greenland shrimp fisheries has been associated with increased discards. Oxygen depletion due to excess organic loading from discards has been recorded in the New Zealand fisheries for hoki (Macruronus novaezelandie) as well as in the North Eastern Atlantic.

Bycatch mortality also affects many non-fish species which are relevant to the functioning of the overall ecosystem. For example, surface and subsurface driftnet and long-line fisheries have serious negative effects on populations of sea birds, e.g. albatrosses and petrels in long-line fisheries in the North Pacific and in the Southern Ocean. High seas drift nets have had a considerable impact on sea birds in the northern Pacific as have gillnets in southwest Greenland, eastern Canada and elsewhere.

Loss of Fishing Gear and Ghost Fishing

Voluntary dumping or loss of fishing gear may lead to ghost fishing. The scale of the impacts of ghost fishing is basically unknown but there are indications that the effects are not negligible. Species affected by discarded gear include not only teleost fish but sea birds, marine mammals and turtles. For example, the incidence of entanglement of marine mammals in floating synthetic debris in the Bering Sea has been related to growing fishing effort and increased use of plastic.

Fowler concludes that entanglement is the principal cause of the current decline in the fur seal population of the Pribilof Islands, accounting for 15% of the mortality of youngsters. As an average, a northern fur seal (Callorhinus ursinus) is expected to encounter 3-25 pieces of net debris along the 800-km yearly migration in the Northeast Pacific. Fish traps, unless made of biodegradable material, contribute to the problem. To illustrate this problem, 31 600 pots were lost in the Bristol Bay crab fishery in a period of two years.

Overall Impacts

The ocean is used for a wide range of human consumptive and non-consumptive uses providing recreation, food, livelihood, energy and pharmaceuticals. It is also used for activities such as transportation, defence, mining, conservation and scientific research. Ample documentation of these uses can be found on the web-based UN Atlas of the Oceans. Non-fishing activities, whether coastal or continental, may have major impacts on the aquatic ecosystems through contamination, habitat modifications and alteration of freshwater flows.

The relevance to Ecosystem Approach to Fisheries (EAF) stems from the fact that most of these impacts will threaten fisheries sustainability, either reducing resource productivity, stability and resilience to fishing, or reducing the quality of the fish as food through contamination. Through land drainage, sewage, river outflow, wind and rainfall, such economic activities as agriculture, manufacturing or chemical industries, incineration of toxic wastes, human settlements, etc., release excess nutrients as well as contaminants (e.g. polychlorinated biphenyls (PCBs), mercury, dioxin), radioactive wastes, oil, antifouling paints (tributyl tin), human pathogens (e.g. cholera, salmonella), plastic and other debris.

Coastal activities, including human settlements and tourism, often result in conversion and destruction of habitats of high relevance to fisheries such as estuaries or coastal wetlands used by fishery resources as reproduction, nursery or feeding areas, reducing fisheries productivity and resilience. Irrigation and production of hydroenergy reduce freshwater inflows into the oceans, resulting in modification or suppression of seasonal floods (as in the Nile), reducing or eliminating key environmental signals and reducing influx of nutrients.

Dams and drying-up of streams, together with mining of gravel and sand and deforestation, modify the quality of habitats and alter or interrupt the reproductive migration processes of many species of marine fish (e.g. salmon, sturgeon). Together, these impacts result in eutrophication, harmful algal blooms, accidental death of animals through entanglement in (or ingestion of) debris; food contamination, human diseases, climate and sea-level changes; UV and temperature changes with impacts on primary productivity and biological habitats (such as coral bleaching) and can lead to total eradication of productive habitats (e.g. in the dried up Aral Sea).

Assessing the scale of fisheries effects relative to other impacts can be difficult, because of confounding and interacting combinations with other anthropogenic effects (e.g. pollution, habitat degradation, climate change) and natural variability of environmental factors. In the early 1970s, pollution and habitat degradation originating from land-based activities were considered to be the main factors of fisheries degradation. More recent publications hold that excessive fishing has become the main destabilising factor of ecosystems, directly, through removals and associated impacts, as well as indirectly, through the aggravation of eutrophication and subsequent oxygen depletion.

While fishing activities are generally fairly well described and their impact on resources have been studied and documented for decades, land-based sources pollution and their impact on the marine ecosystem and on fisheries are very poorly documented. As a consequence, fisheries impacts tend to be the ones more "visible" to the public. Because the fisheries sector is economically and socially weaker than the agricultural or industrial sector, there is a risk that governments and NGOs target it primarily when looking for short-term solutions to the growing environmental problem, raising the issue of intersectoral equity.

The combined impacts from fisheries and land-based activities represent a very significant threat to aquatic ecosystems and to fisheries. The evolution of the Black Sea ecosystem provides an illustration of the problem. The freshwater input from land (346 km^3 per year) is higher than the evaporation (323 km^3 per year) and the bottom is covered with a stratum of anoxic salt water with a high methane and hydrogen sulphide content.

The Danube, Dniestr and Dniepr rivers carry to the Black Sea about 280 km^3 of fresh water per year, representing 85% of the total input of fresh water into that sea. The watershed of the three rivers covers an area (1.4 million km^2) that is 22 times larger than the Black Sea itself, and the Danube alone carries the waste water of 80 million people. At the end of the 1960s, the Black Sea was the most productive area of the Mediterranean, with a high diversity of pelagic and benthic fauna.

After 1970, a very strong modification of the chemical and biological habitat occurred as a consequence of industrial development and intensive agriculture. Inputs of phosphorus and nitrate increased threefold and tenfold respectively while the input of silicate decreased fivefold. This resulted in a significant modification of the structure and functioning of the coastal ecosystem, including a change in dominance of the algal communities from diatoms to small- size dinoflagellates.

At the beginning these modifications appeared favourable for the ecosystem and fisheries. Phytoplankton production and copepod abundance increased and with them the abundance of plankton-feeding fishes. Fishing effort increased, leading initially to an increase in catches from 200 000 tonnes in 1970 to 600 000 tonnes in 1985 but resulting finally in overfishing. The ultimate consequence was an explosion of carnivore jellyfish (*Aurelia aurita* and *Mnemiopsis leidy*) consuming eggs and larvae and occupying the ecological niche formerly occupied by the small pelagic species depleted by overfishing.

The biomass of jellyfish increased from one million tonnes in 1970 to 700 million tonnes in 1985 (about 5 kg/m^2). These jellyfish, having no predators, are a trophic "dead end", and their mortality generates an important bacterial activity and a large quantity of anoxic water near the bottom, reducing further the habitat for fishery resources.

Impact on Diadromous Fish

Diadromous fish, using both the inland and marine environments, illustrate the impact of non-fishery activities on fisheries. Habitat alteration, degradation and outright destruction by non-fishery activities are probably the primary cause of extinction in diadromous fishes. For example, among the diadromous species, the sturgeons are particularly threatened, with 13 of the 15 existing species being endangered, and restoration programmes and restocking will not lead to recovery unless the fundamental sources of decline (in addition to overexploitation), such as pollution and habitat degradation, are addressed.

Many changes in fish stocks can be seen to result from changing of river runoff or influx of a layer of warm water. For example, the collapse of commercial snook fisheries in Florida during the 1950s has been attributed to habitat alteration and reduction of freshwater flow. Major environmental disturbances responsible for the decline of Salmo salar populations include the development of hydroelectric dams, which impede migration to spawning grounds, and acid rain, which degrades quality of nursery grounds.

The poor state of Pacific salmon (Oncorhynchus spp.) is also a consequence of the economic development and exploitation of Northwest Pacific ecosystems, including trapping (furs), mining, timber harvesting, grazing, irrigation, dams, municipal and industrial development, pollution and excessive harvest.

Humans and Marine Mammals

Marine mammals are important predators located towards the apex of the oceanic food chain. They have been heavily depleted in the past and are still, in some cases, threatened with extinction. After decades of protection, many populations are again (or still) very abundant, and their feeding requirements place a very significant demand on ocean productivity. Some countries and stakeholder groups are actively promoting the full protection of these species, questioning our capacity to establish a sustainable regime of exploitation. Other countries consider these species as essential resources and as competitors to humans, pointing to the very large quantities of fish consumed by them.

Non-fishery impacts on resource abundance result in a reduction of the fishery resources available to humankind. While they may not be perceived as such by the fishery sector, they represent de facto a forced, non-negotiated, non-transparent, and potentially non-sustainable allocation of aquatic space and resources to non-fishing activities, often located far inshore (e.g. sources of pollution).

What is an Ecosystem Approach to Fisheries?

The term ecosystem approach to fisheries (EAF) has been adopted in these guidelines to reflect the merging of two different but related and - it is hoped - converging paradigms. The first is that of ecosystem management, which aims to meet its goal of conserving the structure, diversity and functioning of ecosystems through management actions that focus on the biophysical components of ecosystems (e.g. introduction of protected areas). The second is that of fisheries management, which aims to meet the goals of satisfying societal and human needs for food and economic benefits through management actions that focus on the fishing activity and the target resource.

Up until recently, these two paradigms have tended to diverge into two different perspectives, but the concept of sustainable development requires them to converge towards a more holistic approach that balances both human well-being and ecological well-being. EAF is, in effect, a way to implement sustainable development in a fisheries context. It builds on current fisheries management practices and more explicitly recognizes the interdependence between human well-being and ecosystem well-being. EAF emphasizes the need to maintain or improve ecosystem health and productivity to maintain or increase fisheries production for both present and future generations. Of special relevance to these guidelines is the recognition that, in contributing to a convergence of the two paradigms, EAF will be assisting in implementing many of the provisions contained in the FAO Code of Conduct for Responsible Fisheries.

Fishing activities normally target one or several species, known to provide food for consumers and income/livelihood to the fishers. During the past 50 years at least, the dominant fisheries management paradigm has been to maintain the target resource base through various controls on the size and operations of the fishing activity. In these guidelines, we will adopt the term "target resources-oriented management" (TROM) for this paradigm, recognizing that it has been adopted mainly for medium- to large-scale commercial fisheries. In most developing countries (with notable exceptions) and in many developed ones, the activities of the small-scale, multi-species fisheries are undertaken with little intervention beyond development support, or are based on more traditional management systems. The term "current fishery management practices" refers to this global situation, in which TROM is a part.

The depleted state of many of the world's fisheries and the degraded nature of many marine ecosystems have been well documented. Because fisheries have not been managed in a way that contributes positively to sustainable development, the impact on the world's economies and societies will be enormous both now, and probably even more importantly, well into the future. This situation will inevitably contribute to increased poverty, increased inequities and lack of opportunities for many of the world's fishers to make a decent livelihood. Poor management is depriving many regions and states of the potential social and economic benefits of fishing (currently estimated to employ 12.5 million people with about US$40 billion per annum in international trade). Approximately 80–90 million people, most of them in developing countries, depend on fish for their main daily source of protein. The need to reduce the alarming trend of depletion and degradation has been recognized in many international fora, most recently at the World Summit for Sustainable Development (Johannesburg, 2002), which pledged to:

maintain or restore stocks to levels that can produce the maximum sustainable yield with the aim of achieving these goals for depleted stocks on an urgent basis and where possible not later than 2015.

There is obviously a need to improve the approach used in fisheries management so that potential social and economic benefits can be achieved. Conflicts between competing users must be reduced, and fisheries must be accepted by society as responsible users of the marine environment.

Interest in an ecosystem approach to fisheries (EAF) has been motivated by:

— heightened awareness of the importance of interactions among fishery resources and between fishery resources and the ecosystems within which they exist;
— recognition of the wide range of societal objectives for, and values of, fishery resources and marine ecosystems within the context of sustainable development;
— poor performance of current management approaches as witnessed by the poor state of many the world's fisheries; and

— recent advances in science, which highlight knowledge and uncertainties about the functional value of ecosystems to humans (i.e. the goods and services they are capable of providing).

Overall, there is a deeper and broader sense of stewardship in response to increased awareness of the importance of resources and about the current status of fisheries (such as the common occurrence of overfishing, economic waste and adverse impacts on habitat).

In both large- and small-scale fisheries, fishing activities usually affect other components of the ecosystem in which the harvesting is occurring; for example, there is often by-catch of non-targeted species, physical damage to habitats, food-chain effects, or changes to biodiversity. In the context of sustainable development, responsible fisheries management must consider the broader impact of fisheries on the ecosystem as a whole, taking biodiversity into account. The objective is the sustainable use of the whole system, not just a targeted species.

The need for a wider consideration of environmental and ecosystem issues in fisheries has also been acknowledged in many fora, and the principles and aspirations for EAF have been well documented. Although full implementation of agreed principles and aspirations might be difficult at this time, the status quo is not an acceptable option in the light of growing understanding of ecosystems and their uses by society. Progress in implementing EAF is possible, whatever the current approach to managing various types of fisheries. This document elaborates the benefits of EAF and provides practical guidelines for making the changes necessary for an ecosystem approach to marine capture fisheries.

In theory, all aspects of responsible fisheries, as outlined in the FAO Code of Conduct for Responsible Fisheries, can be addressed through EAF. However, the focus of these guidelines is on fisheries management (Article 7) with some coverage of research (Article 11), integration of fisheries into coastal area management (Article 10) and special requirements of developing countries (Article 5). The need to prevent pollution from fishing activities and the impact of polluters on fishing is also included, but was not fully elaborated.

The purpose of EAF can be inferred from many international instruments, reports and scientific publications. Generally speaking, *the purpose of an ecosystem approach to fisheries is to plan, develop and manage fisheries in a manner that addresses the multiple needs and desires of societies, without jeopardizing the options for future generations to benefit from the full range of goods and services provided by marine ecosystems.*

To fulfil this purpose, an EAF should address components of ecosystems within a geographic area in a more holistic manner than is used in the current TROM approach. Doing so will require identifying exploited ecosystems (in their geographic context); their complex nature must be recognized and addressed. An EAF also requires the recognition of many (sometimes competing) societal interests in fisheries and marine ecosystems. Accordingly, this definition follows: *an ecosystem approach to fisheries (EAF) strives to balance diverse societal objectives, by taking account of the knowledge and uncertainties of biotic, abiotic and human components of ecosystems and their interactions and applying an integrated approach to fisheries within ecologically meaningful boundaries.*

EAF is neither inconsistent with, nor a replacement for, current fisheries management approaches (e.g. as described in the FM Guidelines). Rigorously applying TROM approaches (with appropriate emphasis on the precautionary approach and rights-based allocation) would begin to help solve some of the current fisheries problems. Such action in the past could have prevented a large number of present ecosystem problems. Thus, in practice, EAF in the foreseeable future is likely to be developed as an incremental extension of current fisheries management practices.

Principles and Concepts of EAF

EAF addresses a number of concepts, sometimes referred to as "principles" that have been expressed in various instruments and conventions, and in particular in the Code of Conduct for Responsible Fisheries. These principles generally underpin the high-level policy goals assigned to fishery management at a national or regional scale. In brief, recognizing that fisheries have the potential to alter the structure, biodiversity and productivity of marine ecosystems, and that natural resources should not be allowed to decrease below their level of maximum productivity, fisheries management under EAF should respect the following principles:

- fisheries should be managed to limit their impact on the ecosystem to the extent possible;
- ecological relationships between harvested, dependent and associated species should be maintained;

— management measures should be compatible across the entire distribution of the resource (across jurisdictions and management plans);
— the precautionary approach should be applied because the knowledge on ecosystems is incomplete; and
— governance should ensure both human and ecosystem well-being and equity.

Making EAF operational

There is considerable agreement on the underlying principles of EAF, and on their implications for policy. There is also consensus among academics, scientists, fishery advisers and many non-governmental organizations (NGOs) on the essential elements of an ecosystem approach to fisheries. However, to implement EAF it is necessary to translate the principles into operational objectives and action.

Translation of principles into high-level policy goals is relatively simple in terms of wording and definitions. Policy goals will usually reflect the overarching principles outlined in relevant domestic legislation, regional agreements and international agreements of various kinds. There should also be some societal agreement on the degree to which it is acceptable for fisheries and other users to alter these "characteristics".

Translation of policy into action is more important, but it is probably the most difficult step in the implementation of principles. At the outset, all stakeholders must recognize the existence of a hierarchy of issues together with related objectives, indicators and performance measures. Without this recognition, EAF will simply remain an important concept, but will not really be useful in day-to-day fisheries management.

The aim of these guidelines is to translate the high-level policy goals into action by:

— identifying broad objectives relevant to the fishery (or area) in question;
— further breaking these objectives down into smaller priority issues and sub-issues that can be addressed by management measures;
— setting operational objectives;
— developing indicators and reference points;
— developing decision rules on how the management measures are to be applied; and

— monitoring and evaluating performance.

It is not possible to be prescriptive on these sub-issues because they will obviously vary among fisheries. However, it is important to consider all the economic, social and environmental aspects of fisheries so that an important issue or sub-issue is not overlooked.

Any advice or guidelines then need to take into consideration the differences between developed and developing countries or types of jurisdiction, the availability of handbooks and manuals as well as technical protocols (e.g. to develop indicators), training of scientists and managers, etc.

Moving towards EAF management

In this section, the topics covered in the FM Guidelines are considered sequentially in terms of the limitations of current fisheries management practice and what would be required to fully implement EAF, noting that current management practice frequently falls short of TROM requirements and paradigms. As applied in the FM Guidelines, it is useful to categorize the different aspects of EAF into (i) the fisheries management process, (ii) the biological and environmental concepts and constraints, (iii) technological considerations, (iv) the social and economic dimensions, (v) institutional concepts and functions, (vi) time scales in the fisheries management process and (vii) the precautionary approach. Based on the increased emphasis of the importance of fish and fisheries to developing countries, a further category, (viii) special requirements of developing countries, has been added.

The main limitation of most current fisheries management is that it fails to effectively take into account the interactions that occur between fisheries and ecosystems and the fact that both are affected by natural long-term variability as well as non-fishery extractive and polluting activities.

The Fisheries Management Process

The current fisheries management practice of planning, setting objectives, implementing strategies and measures to meet the objectives, as well as monitoring and assessing performance, if conducted to a satisfactory standard, will still provide a sound basis for implementing EAF. However, recognizing the broader economic and social interests of stakeholders under EAF, the setting of economic and social objectives will need a broader consideration of ecological values and constraints than is currently the case.

This will require a broader stakeholder base, increased participation and improved linkages of fisheries management with coastal/ocean planning and integrated coastal zone management activities.

Biological and Environmental Concepts and Constraints

Marine capture fisheries affect the environment directly (e.g. removal of target and non-target species, habitat change) and indirectly (e.g. changing biological interactions). Similarly, changes in the environment (e.g. climate, agricultural practices and pollution) affect fisheries.

TROM is based on the paradigm that the productivity of marine systems and the level of harvest for any target are limited. It may refer to non-target species, associated and dependent species but, in general, it does not sufficiently recognize the potential direct and indirect effects of fishing on the dynamics of the ecosystem, the conditions under which its productivity can be maintained and the existence of other societal values and uses. TROM is often based on a management unit (e.g. species, gear and jurisdiction) that takes little account of the ecosystem structure or boundaries in which it is operating.

EAF is based on the same "paradigm of limits" as TROM. It recognizes that our ability to predict ecosystem behaviour is inadequate, and accepts that all ecosystems have limits that, when exceeded, can result in major ecosystem change – possibly irreversibly. Maintaining biological diversity is regarded as being of major importance to ecosystem functioning and productive fisheries, as well as providing flexibility for future uses. Current management practices tend to give insufficient recognition to the fact that many components are intrinsically linked in the system in a complex flow of material, energy and information.

There have been many attempts to define an ecosystem. A fundamental principle is that ecosystems are one in a hierarchy of biological organizations in which the integrated whole is more than the sum of the parts (e.g. cells, organisms, ecosystems and biosphere) and are comprised of both living plants and animals (including man) as well as non-living or abiotic structures. They can be defined at many scales, for example from a boulder on a reef to an entire ocean. They can, therefore, overlap or be nested together. Ecosystems are usually spatially defined (i.e. they are sufficiently different from adjacent areas to be recognized as a functional unit) but most of them have no fixed boundaries, especially within the marine environment, and they exchange matter and information with neighbouring ecosystems. However to be able to implement EAF at an operational level, delineation of the "boundaries" is required and can be achieved by a sensible consensus based on proposed EAF objectives.

Technological Considerations

EAF seeks to build on conventional fishery management measures to regulate fishing mortality through the use of input controls, output controls and technical measures (including spatial measures) by broadening the approaches to include other measures such as modifying populations by restocking or culling, where appropriate and effective. Similarly, habitat restoration and MPAs will need to be considered both in the context of facilitating fishing activity or enhancing the populations of target species as well as protecting biodiversity and providing broader benefits to the system as a whole.

Gear modifications, such as those used to selectively harvest the target species and minimize unwanted by-catch, including protected species, for example turtle exclusion devices (TEDs) and by-catch reduction devices (BRDs), will take on increasing importance as ecological objectives are broadened within the context of EAF. The impact of some fishing gear and methods on the bottom habitat (biotic and abiotic) can often have a negative effect on the ecosystem. There is limited knowledge about this impact, however, and more research is required to examine the extent of the impact of various gear. For gear known to produce serious impacts, the introduction of restrictions may be necessary and, where possible, new technologies that mitigate any negative impact will need to be developed.

Fishing operations may also cause other negative impacts to the environment, such as continued fishing by lost gear ("ghost fishing"), emission of exhaust gas with dangerous substances to the atmosphere and pollution from oily waste, litter and fish waste. Minimizing such impacts will require development and successful introduction of alternative cost-effective technologies and fishing practices.

Many ecosystems, especially those in coastal waters, are impacted not only by fisheries, but also by other users, including upstream land-based activities. In these cases, many of the broader measures will be the responsibility of other agencies. Fisheries managers will need to take a proactive approach so that the appropriate authorities recognize fisheries as an important stakeholder in these ecosystems.

Social and Economic Dimensions

Current fisheries management often focuses on a limited set of societal goals and objectives for achieving economic and social benefits from fishing. However, as the overarching goal of EAF is to implement sustainable development, the shift to EAF will entail the recognition of the wider economic, social and cultural benefits that can be derived from fisheries resources and the ecosystems in which they occur. The identification of the various direct and indirect uses and users of these resources and ecosystems is a necessary first step to attain a good understanding of the full range of potential benefits. While many of these benefits may be amenable to quantitative assessments, some are not, and their value can be described only in qualitative terms. Multi-criteria decision-making techniques may be applied to create aggregate indices that encapsulate both quantitative and qualitative ecological, economic, social and cultural considerations.

The quantitative valuation of marine ecosystem goods and services can be based on the concept of total economic value (i.e. use and non-use value). Many ecosystem goods and services are not traded, and therefore need to be valued through means other than market prices. While various approaches have been developed to undertake such valuations, they pose particular difficulties in the measurement of non-use values, especially current or future (potential) values associated with resources which rely merely on continued existence of the resource and are unrelated to use (e.g. conservation of some endangered species). The relative weights given to use and non-use values by different groups, not just within countries but also between countries, can give rise to diverging views on whether specific fishing practices should be modified or cease entirely.

The consideration of a broader range of ecosystem goods and services necessarily implies the need of addressing a wider range of trade-offs between different uses, non-uses, and user groups. In view of the higher complexity of EAF and limited ability to predict changes in the future flow of ecosystem goods and services, valuation has to take uncertainties and risks explicitly into account.

Ecosystem considerations have been part of the fishery perspective of many traditional fishing communities for long periods in different parts of the world. Nevertheless, overcapacity, overfishing and destructive practices have also occurred in many small-scale fisheries. EAF provides a framework within which traditional fisheries management practices can be recognized and strengthened to address some of these problems. EAF is better suited than TROM to handle impacts arising from destructive fishing practices, habitat degradation and pollution, and to use traditional ecological knowledge about fish and their habitats. EAF must, however, take into account the dependence of artisanal and small-scale fishing communities on fishing for their life, livelihoods and food security.

Institutional Concepts and Functions

One of the implications of implementing EAF is an expansion of stakeholder groups and sectoral linkages. This may have substantial impact on institutional structure and process, in terms either of creating new structures or strengthening existing institutional collaboration. Division of responsibilities within governments and differing priorities among different economic sectors are impediments to be overcome in order to implement an ecosystem approach to fisheries. An effective ecosystem approach will depend on better institutional coordination (e.g. between ministries). This will require clarification of roles and responsibilities, improved coordination and integration across government and other users and more accountability across all stakeholder groups. A greater emphasis on planning at a range of geographical levels that involves all relevant stakeholders will be required and will involve a much more collaborative approach and sharing of information. The magnitude of this task should not be underestimated, and a global acceptance of the benefits of this approach is needed for it to succeed.

In many cases, fisheries are currently managed by an agency with narrow legislation and objectives pertaining to the harvesting of only the target species without due regard to other uses/users in the area of the fishery or its impact on the ecosystem. Many laws and regulations may need to be changed to incorporate EAF. Management units may need to be redefined geographically or, at the very least, coordinated within a larger-scale planning process. This will be particularly important where natural and operational boundaries straddle jurisdictions and management plans, and where the indirect effects of fisheries are manifested elsewhere.

In most countries, EAF will require considerable capacity building. This will include improving understanding of ecosystem structures and functions; training managers and regulators to deal with a broader range of options and trade-offs, conflicts, rights and regulations; and enhancing stakeholder capacity to participate. This may be achieved, at least in part, by mobilizing and linking with existing institutions.

Time Scales

The FM Guidelines recognize three time scales of immediate relevance to the fisheries management process – a policy cycle of about 5 years, a fishery management planning and strategy cycle of 3–5 years and a shorter cycle of management implementation and review at an operational level, usually occurring annually. These will also apply to EAF, although the coordination necessary to achieve EAF may mean that progress is slower in some more complicated areas. Longer time scales will need to be considered when dealing with issues such as climate change or the well-being of future fisheries generations.

Precautionary Approach

Under EAF, the precautionary approach gains even greater significance, because it is expected that uncertainty will be much greater than under TROM. Application of the principle specified in the FAO Technical guidelines on the precautionary approach to capture fisheries and species introductions that "where there are threats of serious irreversible damage, lack of full scientific certainty shall not be used as a reason for postponing cost-effective measures to prevent environmental degradation" should result in conservative management action being taken until more is known about ecosystem structures and functions. Under EAF, as outlined in the above-mentioned publication, the principle is much broader than just environmental degradation, and applies to any undesirable outcome (ecological, social or economic); it should also be applied in all stages of the management process.

Special Requirements of Developing Countries

The challenge to implement improved fisheries management is stretching national systems and capacity in most countries, and especially in the developing world. Implementing EAF could add a significant additional burden, and the challenge may be particularly formidable in small-scale fisheries, where the difficulty and costs of the transition to effective management may outweigh the available capacity and short-term economic benefits derived from it. Particular problems are likely to be encountered in regions where poverty is widespread, alternatives to fishing are scarce or non-existent, and where the traditional systems have broken down. In such situations, the short-term economic necessities, at both national and local levels, may be too overwhelming for serious consideration of change even when the long-term benefits are apparent.

The particular problems being faced by developing countries in implementing the Code of Conduct and EAF, and the role of the international community in assisting them, have already been recognized in major international instruments. In particular, Article 5 of the Code of Conduct, Special Requirements of Developing Countries, states:

> In order to achieve the objectives of this Code and to support its effective implementation, countries, relevant international organizations, whether governmental or non-governmental, and financial institutions should give full recognition to the special circumstances and requirements of developing countries, including in particular the least-developed among them, and small island developing countries... especially in the areas of financial and technical assistance, technology transfer, training and scientific cooperation and in enhancing their ability to develop their own fisheries as well as to participate in high seas fisheries, including access to such fisheries.

Paragraph 30c of the Plan of Implementation of the World Summit on Sustainable Development drew attention to Article 5 of the Code of Conduct, and the 2001 Reykjavik Declaration affirmed:

> Our determination to strengthen international cooperation with the aim of supporting developing countries in incorporating ecosystem considerations into fisheries management, in particular in building their expertise through education and training for collecting and processing the biological, oceanographic, ecological and fisheries data needed for designing, implementing and upgrading management strategies.

Greater attention needs to be given to fulfilling these requirements if the developing countries as a whole are to be able to make progress in implementing the growing number of agreements and instruments aimed at fisheries and fishery resources, as these countries simultaneously struggle with pressing fundamental socio-economic issues such as food security, health and access to other basic necessities.

To mobilize more national resources, every opportunity should be taken to raise awareness and facilitate the use of EAF in all relevant cases. To justify using public financial resources, the many benefits that can be derived from the approach, not just those for the fishery sector, need to be highlighted. Emphasis also has to be placed on the existence of potentially high returns from improved management in order to mobilize support from international financial institutions.

The following issues are will need to be addressed to assist the implementation of EAF in developing countries:

— *Adaptation to capacity-poor situations.* Efforts are needed to tailor EAF to the capacity available in developing countries and small-scale fisheries, focusing on data-poor situations and providing appropriate models and methods for such situations. In addition, participatory and adaptive approaches will need to be developed, drawing on existing traditional rights and management systems whenever possible. There may also be advantages in integrating fisheries management into coastal area management where it could benefit from economies of scale and the existing networks for participation.

— *Financial policies.* International financial agencies and mechanisms as well as national development banks will need to facilitate and contribute to the finances necessary to take action on EAF. In appropriate cases, mechanisms could be established to recoup this funding through proper capture of the economic rent generated by better management (including payments for rights). Investing in disinvestment should also be seriously considered in suitable cases.

— *Aid and technical assistance.* Fish are a global commodity needed in the rich areas of the world as well as in the poorest, and building up a long-term national and regional institutional capacity to manage resources sustainably should be considered a global "duty". International financial institutions should adopt measures to assist developing countries to restore and manage their fisheries to facilitate food security and livelihood opportunities for impoverished coastal communities. Priority should be given to the least developed and food-deficit countries.

Data and Information Requirements

Data and information are the basis of good management. They underpin all stages in the EAF management process including formulating policy, developing management plans, and evaluating progress and updating policy and plans to provide for continuous improvement. As pointed out in FM Guidelines, although the data and information required for each of these stages overlap, the processes are distinct, occur on different time scales and require information at different levels of detail and aggregation. The guidelines in this Supplement do not re-iterate many of the important points concerning data collection and analyses already stated in the FM Guidelines, but attempt to show instead where EAF will require a broadening of data, analyses and information provision.

Because EAF is a broadening of current fisheries management practices, the data and information needs will by necessity be broader. However, it is important to stress that immediate action should be based, as much as possible, on data and information that already exist. In some countries, much of the information will already be available in reports and statistics from various research institutes, agencies and ministries. In others, EAF will have to be based on comparatively fewer data. However, in these cases there is often extensive traditional knowledge about the ecosystem and the fishery, which can be extremely useful if collected and validated from interviews with local fishermen and other stakeholders. In all cases, information about the local situation should be complemented by information from ecologically similar situations elsewhere.

Policy Formulation

Policy development will be informed by broad information on the role that fisheries play in terms of the regional, national and local economy and social setting. As in TROM and other fishery management responses, information should be collected about the stakeholders, economic factors related to the fishery, details on costs and benefits, role in providing employment or livelihood, alternative sources of employment and livelihoods, status of access to or ownership of the resource, institutions currently involved in planning and decision-making, along with a historical perspective of the fishery and its stakeholders. Under EAF, similar knowledge of alternative uses and users of the resources within the ecosystem will be required, and a better understanding of the many interactions that occur within the system is fundamental. A fishery will often affect species whose distribution extends beyond the distribution area of the fishery. Other users should also be informed by the fishery sector on the role fisheries play in the broader social and economical setting and on how any actions may affect other stakeholders.

Developing Management Plans

Formulating management plans is an important component of implementing EAF. To the extent possible, plans must be based on an understanding of a broad background of knowledge, although a lack of data or uncertainty about the impact of the fishery should not be used as an argument for delaying the formulation of an EAF management plan. Only in situations where the existing information is insufficient to decide whether a potentially important impact does actually take place will it be necessary to collect and analyse additional data.

The information that feeds into a fishery management plan should include:

— the area of operation of the fishery and its jurisdiction;

— the various stakeholders;

— the gear and vessel types to be employed in the fishery;

— the history, management and socio-economic importance of the fishery;

— if possible, the distribution area of the most important commercial species in the catch (preferably as a map);

— relevant information about the life histories of these species;

— the effects of the fishery on the recruitment, abundance, spatial distribution and age or size structure of the target species, as far as possible;

— any available monitoring data; and

— any management procedures already in place, with descriptions and a performance evaluation.

In addition to these TROM requirements, the potential direct and indirect effects of the fishery on species and habitats will also need to be described. Ideally, the information should consider the following, but if this is not possible, at least a comment about the following should be included:

— the critical habitats that may be affected and the potential direct and indirect impacts of the fishery on these habitats;

— the species composition of both the retained and non-retained by-catch and the potential effects of additional fisheries-generated mortality on affected populations;

— the likely amounts of discards produced by the fishery and the importance of these discards for potential scavengers;

— the potential amounts of litter produced by the fishery and the possible effects of lost or abandoned gear on fish and other biota;

— the ecosystem within which the fishery takes place including the impact of other anthropogenic activities such as releases of nutrients and contaminants;

— the major biological interactions in which the harvested species participate and the potential effects of fisheries on these interactions. Particular efforts should be made to identify possible interactions with critical species, with forage species important for transfer of energy in the food chain, and with habitat structuring species such as coral;

— the impact of fishing on life history traits, such as age and size of first maturity and possible effects of the fishery on the genetic diversity of affected populations;

— the legal framework and extent to which the effects generated by the fishery would comply with national regulations and with international law and agreements related to nature conservation with consideration for endangered species; and

— the possible management measures to reduce adverse environmental impacts.

The guidelines stress the need to translate policy goals and broad fishery objectives into operational objectives in order to implement EAF. The process also needs to be informed by the best available scientific advice so that, firstly, all the issues relevant to a particular fishery have been covered and secondly, that all alternative objectives, indicators and reference points can be assessed.

Monitoring, Implementing and Performance Reviews

The setting of operational objectives and indicators will identify what information will need to be routinely collected in order to feed into the decision-making process, as well as the short-term (annual) and long-term (3–5years) reviews and assessments of management performance. The indicators that are developed may vary from fishery to fishery, depending on the main issues identified for a particular fishery. However, many fisheries will have a basic set of common issues, objectives, indicators for which data and information will be required. These will cover the ecological (including the fisheries resources), economic and social dimensions of sustainable development.

Uncertainty and the Role of Research

Given the complexity of the ecosystems in which fisheries operate and the dynamic nature of the myriad of interactions that can occur, science (in its broadest sense word including biologists, mathematicians, sociologists, economists and technologists working in collaboration with stakeholders) cannot possibly hope to deliver on all the information required. There is an obvious need for more ecosystem information, for better information on social and ecological implications, for an understanding of the management process itself (including the provision of information in decision support systems) and for monitoring and assessment methods.

Management Measures and Approaches

The measures available to managers to adopt an EAF will, at least in the short term, be an extension of those conventionally used in TROM. Thus the range of input and output controls and technical measures (including spatial measures) used to regulate fishing mortality remain highly relevant; but these controls will need to be considered in a broader context. This means recognizing that the range of measures chosen should not only address a series of target species concerns, but should also enhance ecosystem health and integrity. Managers should consider as far as possible a coherent mix of approaches that takes account of the interdependencies and functioning of the ecosystem. Apart from managing the direct effects of fishing activity, fishery managers will need to be aware of other measures that are available for managing populations (e.g. restocking and culling). Similarly, habitats may be modified to enhance the populations of target species or to restore degraded areas.

While population and habitat manipulation may lie partly within the remit of fishery management bodies, there are many other issues, generally within the competence of other agencies, that concern fisheries managers. These may be highly relevant in an EAF context; they include such issues as the impact associated with human activities on land and sea leading to habitat destruction, eutrophication, contaminants, CO_2 emissions, litter, accidental introduction of exotic species through ballast water, etc. Fishery managers should be proactive in these circumstances to ensure that the appropriate authorities include all those involved in fisheries as important stakeholders in management planning and decision-making.

Options to Manage Fishing

Technical measures

Gear modifications that improve selectivity

Most fishing gear affects marine life in one way or another. One major impact is that gear is used to remove the larger fish from a population and thus to change the size composition of the targeted species. In many fisheries, the gear also has an impact on non-target organisms. They are captured as well, and this by-catch is frequently discarded because of its low economic value, prohibitions on landing or space limitations on board the vessel. The consequences for the ecosystem can be severe. For example, discarding by-catch can often change the trophic structure of entire ecosystems with the encouragement of scavengers, as is seen in many shrimp fisheries around the world. Size selective harvesting can, under some circumstances, lead to genetic changes in affected populations, such as changes in growth and in size and age at first maturity. Under EAF, these effects need to be considered more seriously.

Size selectivity of target species. Mesh size restrictions can be a useful measure to avoid capturing individuals of target species in the immature stages, but they have limitations in multi-species fisheries. When organisms of different shapes and sizes occur on the same fishing ground, immature individuals of a co-occurring larger species might still be captured.

When considering introduction of mesh size regulation in a trawl fishery, it is also important to consider the survival rate of the organisms that escape through codend meshes. If mortality is high, the anticipated benefit of larger meshes may not be achieved. Selectivity can be improved through a variety of methods other than mesh size, including the use of square mesh, sorting grids and other devices which enable the unwanted portion of the catch to escape.

Non-target species selectivity. Tools that reduce capture of non-target species in fisheries are known as by-catch reduction devices. Some successful examples include:

— turtle excluder devices (TEDs);
— sorting grids that assist in allowing unwanted by-catch to escape;
— circle hooks and blue dye baits that reduce incidental capture of turtles in longline fishing;

— scaring lines positioned above a longline gear during setting, thawed bait, night setting with minimum ship light, weighting the line, underwater setting, prohibition of dumping offal during setting to reduce catching seabirds;

— acoustic pingers to distract marine mammals from becoming entangled in gillnets; and

— modified operational methods and gear modifications that avoid capture of dolphin while purse seining for tuna.

All of these measures have proved to be very effective in different fisheries around the world and there are several examples where there have been economic benefits as well as large ecological benefits, e.g. in the Caribbean trap fisheries, in the Alaskan ground fish fishery and in tropical shrimp fisheries in Australia.

Other gear issues

When fishing gear like gillnets and traps/pots are lost during fishing operations, they may continue to capture fish for several weeks, months or even years, depending on the depth and prevailing environmental conditions (light level, temperature, current speed, etc). This "ghost fishing" can be partially limited by using biodegradable materials or some means to disable the gear, through increased effort to avoid losing them, or by facilitating the quick recovery of lost nets. In some areas, active campaigns are undertaken to "sweep" periodically for lost nets in known gillnet fishing grounds.

Spatial and temporal controls on fishing

Fishing mortality can be modified by restricting fishing activity to certain times or seasons, or by restricting fishing in particular areas. Such measures can be used to reduce the mortality rate of individuals of either target or non-target species in vulnerable life stages. Where stocks are shared by more than one country, the closures – like other management measures – must be coordinated.

The selective reduction of fishing mortality rate on both target and non-target species generally reduces both the direct and indirect effects of fishing on the ecosystem. Closures may be used to protect critical habitats where fishing activity would otherwise cause damage to the physical structures supporting the ecosystem. They may also help to reduce mechanical disturbance to the benthos and facilitate the establishment of more stable and structured communities.

One form of closure is that of marine protected areas (MPA)s. MPAs range from "no take" to planned "multiple-use" areas. MPAs are often designated for non-fishery objectives, but they can produce considerable benefits for fisheries. MPAs can protect sedentary species, allow a proportion of the stock to remain free of the genetic selective effects of fishing, and may act as refuges for the accumulation of spawning biomass from which replenishment of surrounding fished areas can occur, either through out-migration of fish or dispersal of juveniles. This latter benefit has yet to be demonstrated unequivocally for a range of locations, and may be site-specific.

Commonly, spatial and temporal closures have been established in the context of specific target stocks or fisheries, and it is not unusual for a very large variety of such ad hoc measures to occur in a single ecosystem. While such an approach may have its benefits, there may be advantages in a more systematic scheme where consideration is given to a coordinated attempt to protect a range of habitats and species on a scale which is relevant to the ecosystem concerned. This requires a synthesis of the current understanding of the important elements of ecosystems and an evaluation of the potential benefits.

It is important to include an evaluation of the overall effect of a closure based on the biology of the species concerned and the nature of the fishery. The success of spatial and temporal closures can be limited if their effect is merely to displace fishing activity and increase mortality of other species or life stages elsewhere. Species that are mobile and move between the protected and non-protected areas may, in fact, gain little protection.

Area closures that permit some fishing may require a large enforcement effort and can therefore be costly. Allowing certain categories of fishing activity can also create loopholes which undermine the intentions of the closure. Management authorities need to consider the likely degree of compliance and enforcement costs in establishing closures, although the advent of vessel monitoring systems (VMS) makes area-based management more enforceable in some regions of the world.

Control of the impact from fishing gear on habitats

Fishing gear that touches or scrapes the bottom during fishing operations is likely to produce negative impact on the biotic and abiotic habitats. Because only limited knowledge exists about the long-term effect of such impact, a precautionary approach is recommended in the use of high-impact fishing methods in critical habitats. Use of towed gear with reduced bottom contact is a technical option in such areas. Prohibition of certain gear in some habitats is another, e.g. trawling in coral reef and seagrass areas. A third option is to replace a high-impact fishing method with one with less impact on the bottom, e.g. trapping, longlining or gillnetting.

Energy efficiency and pollution

Many modern fishing vessels use fossil fuel for propulsion, for operating the fishing gear and for the preservation and processing of the catch. The impact of exhaust gas emission of dangerous substances, including CO_2, has been fully recognized, and technological innovations that reduce such emissions are encouraged. Energy optimization can be achieved through improved efficiency of fishing gear as well as through improved management that lead to less fishing effort being required.

Input (effort) and output (catch) control

Controlling overall fishing mortality

The direct effects of fisheries on marine ecosystems are to increase fishing mortality rate among target and non-target species and to affect habitat. The fishery management methods that are used to control fishing mortality are often referred to as input and output controls. Input controls apply to *capacity* (which is closely related to the fishing mortality a fishing fleet could generate if the entire fleet were to fish full time) and *effort* (which is the actual amount of fishing activity). Output controls apply to the catch that results from the fishing effort. Well-known fisheries models are used to relate both catch and fishing effort to fishing mortality.

Capacity limitation seeks to restrict the total size of the fleet, thus reducing both fishing mortality and the pressures on decision-makers to allow higher fishing mortality. Capacity controls have the potential to reduce fishing mortality on entire species complexes in exactly the same manner as effort or spatial/temporal access limitations.

Effort limitation seeks to restrict the fishing activity of fleets and hence reduce fishing mortality. Because this operates at the fleet level, there will be a reduction in mortality among all species involved in the fishery, and this may be advantageous when dealing with multi-species fisheries. Although there is a considerable difference in the likely social and economic effects of different effort limitation regimes, the net effect of reducing the amount of fishing will produce benefits for the ecosystem, provided the continual improvement in efficiency ("effort creep") does not cancel out the effect over time.

In current fisheries practices, the main limitations of any of these controls are that they do not directly constrain the fleet from targeting and depleting an individual stock. From an EAF viewpoint, these input controls have the virtue of restricting the overall pressure on the ecosystem, thus offering the potential of limiting negative impacts. However, there is also considerable danger that fishing mortality will steadily increase if increasing efficiency is not monitored and controlled. While increases in efficiency, if unchecked, will increase the fishing mortality in the target and by-catch species, some technological progress such as development of echo-sounders and satellite navigation may also enable fishermen to direct more of their effort towards the target species and thus diminish the impact on non-target species.

Catch controls

Catch controls in the form of catch limitations are aimed at directly reducing fishing mortality on target species. If complemented with by-catch controls (such as quotas) they have the potential to protect associated species. They have proven successful in some cases, including in multi-species fisheries, but have sometimes also led to undesirable outcomes (high-grading, increased discarding, etc.). In terms of an EAF, however, in a mixed-species fishery, consideration needs to be given to the different vulnerabilities and productivity of the various species. It will be necessary to implement a set of consistent catch limits across the range of target and by-catch species to reflect these differences and addresse desired ecosystem related objectives (such as maintaining food webs). Catch limits for target species may need to be modified to control catches of more vulnerable species.

Ecosystem Manipulation

In some situations, technology and understanding of marine ecosystems have advanced to the point where ecosystems may be manipulated to achieve societal objectives that include conservation and restoration. Such manipulation (in the form of, for example, stock enhancement, culling or habitat restoration) may be an attractive option to mitigate negative impact from the past (like overfishing or habitat destruction). However, mitigation is rarely completely effective, carries with it some risk of unexpected consequences; it may also be costly. There is still little experience with successful ecosystem manipulation, and knowledge is insufficient to allow for sound prognoses. Avoiding the causes of the problem in the first place is a much more desirable approach.

Habitat modifications

Preventing habitat degradation. Habitat preservation in marine fisheries is the key to EAF, because it underpins the health of exploited ecosystems. Managers need measures to prevent damage to habitats, to restore damage where it has occurred and to increase habitat where required. Such measures must be in harmony with other ecosystem functions. Various types of fishing pose threats to the integrity of the habitats that support fisheries resources and other components of the ecosystem. Apart from notorious practices such as using dynamite and fishing with poison, already widely outlawed, several other practices may result in physical and biological damage to the seafloor. The different measures needed to reduce such impacts include:

— prohibition of destructive fishing methods in ecologically sensitive habitats (such as seagrass beds);

— prohibition of intentional cleaning of the seafloor to facilitate fishing; and

— reduction of the intensity of fishing in some fishing grounds to ensure that non-target, habitat-forming species are not reduced below acceptable levels.

Providing additional habitat. In situations were it is evident that insufficient habitat is available to support species of interest or concern, additional habitat can be created in two ways. The first measure applies where habitat has been damaged or lost and involves re-establishing mangroves, seagrasses and coral reefs. Such rehabilitation programmes should not be implemented unless the problems causing the damage in the first place have been adequately addressed. The primary objective is to re-create the physical structure needed to provide shelter for animals and a substrate for forage organisms. Ideally, rehabilitation programmes should aim to increase biodiversity, so they should aim to be multi-species rather than monospecific enhancements. In some cases, simply providing the conditions necessary for survival of propagules (coral larvae, seagrass seeds) arriving from nearby areas of habitat will result in restoration of habitats. Because many species of fish use different habitats as a continuum during their development, restoring only some habitats may not achieve the full potential of a rehabilitation programme to improve productivity or biodiversity.

The second method is to construct artificial habitat. Well-designed and -located artificial habitats have the potential to improve production by increasing the settlement success of juveniles in years of abundant seed supply (e.g. larvae). Artificial habitats may also play an integral part in a restocking or stock enhancement programme by permitting a larger number of animals to be released. However, care needs to taken to ensure that the new habitat does not redistribute fish in a way that makes them more vulnerable to overfishing. Artificial habitats may also become a navigation hazard, pollute the ecosystem or disrupt its structure and function. Problems can also occur when the artificial habitats are not robust enough to prevent them from breaking up during storms and littering the seashore.

Decisions to increase the amount of structural habitat will involve choices about the relative value of different components of the ecosystem (habitats and species), because creation of one habitat will be at the expense of another. Artificial habitats are also expensive to construct and it may be more effective to protect the existing natural and renewable forms of fish shelters, such as seagrass beds.

Population manipulation

Restocking and stock enhancement. Target species that have been heavily over-exploited in some fisheries ecosystems can potentially be restored by releasing cultured juveniles to rebuild the spawning biomass, and then protecting the released animals, the remnant wild stock and the progeny until the population increases to the desired level. This process is known as restocking, and differs from stock enhancement. The former aims to rebuild the stock back up to viable levels, while the latter supplies additional stock to harvest. However, as there are often high costs involved in restocking programmes, careful analysis is needed to determine whether the goals of rebuilding stocks can be achieved by other management measures. In general, restocking should be considered only when other forms of management are incapable of restoring populations to acceptable levels, and it should be coupled with controlled fishing capacity and reduced overfishing. If restocking is needed, and the species is part of a mixed fishery that need not otherwise be closed, restocking can be carried out in MPAs.

To reduce the risks of adverse effects on remnant wild individuals of the same species or other species in the ecosystem, restocking programmes must incorporate: (i) hatchery procedures that prevent loss of genetic diversity by guarding against inbreeding and selective breeding and (ii) quarantine protocols that prevent the transfer of pathogens from cultured animals to the wild.

Where managers wish to increase the yields of particular species from ecosystems, release of cultured juveniles in "stock enhancement" can sometimes be used to manipulate population levels. This process aims to overcome recruitment limitation, which occurs when the natural supply of juveniles falls short of the ability of the habitat to support the desired stock level. As with restocking programmes, careless hatchery practices could also result in the release of individuals unfit for survival in the wild, modification of genetic diversity and the introduction of diseases.

Factors to be considered in determining the benefits and costs of stock enhancement programmes include: (i) the need to minimize production of hatchery-reared juveniles by optimizing the scope for natural replenishment by wild stocks, (ii) the abundance of predators and prey at proposed release sites, and (iii) the need for independent assessments to determine whether the enhancement programme is achieving its goals and whether it is having adverse effects on the ecosystem. It may also be necessary to provide additional habitat to support the increased numbers of enhanced species.

Culling. This measure usually aims to reduce the abundance of predators or species that compete for the same trophic resources, in order to increase the yields of target species or to maintain the balance of the trophic structure. However, such food-web manipulation needs to be carried out with caution to ensure that it produces only the desired effect and does not result in unwanted changes in abundance of other important components of the ecosystem or threaten the survival of the species culled. An adaptive approach is needed, which may benefit from planned experimentation in some cases. Consideration should first be given to the rebuilding of target species populations through other, more conventional, fisheries management measures. Large-scale culling should be conducted only after the full implications of the manipulation have been thoroughly investigated.

Intentional introductions. Although new fisheries can be created by introducing species, there is a high risk of causing detrimental changes in coastal ecosystems. A precautionary approach is needed here, but this does not mean that the measure should never be considered. Some introductions of marine species have resulted in social and economic benefits with no apparent impacts on other components of the ecosystem. Fisheries for trochus in the Pacific and scallops in China are good examples.

A comprehensive risk assessment should be undertaken before considering the creation of new fisheries based on introduced species so as to understand the benefits and consequences of such measures. Steps to be undertaken in a risk assessment should include a detailed understanding of issues such as the trophic level of the species, reproductive potential and requirements, interactions with other species, introduction of pathogens and parasites, and effects on demand for and supply of other species.

Rights-based Management Approaches

The dangers and consequences of allowing open access to fisheries are now well understood, where the different options for limiting access and their properties are also described. The Code of Conduct stipulates:

> "States should develop, as appropriate, institutional and legal frameworks in order to... govern access to them (coastal resources) taking into account the rights of coastal fishing communities".

A well-defined and appropriate system of access rights in a fishery produces many essential benefits, most importantly ensuring that fishing effort is commensurate with the productivity of the resource and providing the fishers and fishing communities with longer-term security that enables and encourages them to view the fishery resources as an asset to be conserved and treated responsibly.

There are several different types of use rights. Territorial use rights (TURFs) assign rights to fish to individuals or groups in certain localities. Limited-entry systems allow only a certain number of individuals or vessels to take part in a fishery, with entry being granted by way of a license or other form of permit. Alternatively, entry may be regulated through a system of effort rights (input rights) or by setting catch controls (output rights), where the total allowable catch (TAC) is split into quotas and the quotas allocated to authorized users.

Each type of use right has its own properties, advantages and disadvantages, and the ecological, social, economic and political environment varies from place to place and fishery to fishery. Therefore, no single system of use rights will work under all circumstances. It is necessary to devise the system that best suits the general objectives and context for each case, and this system may well include two or more types of use rights within a single fishery or geographic area. For example, a fishery that includes artisanal and commercial fishers could make use of TURFs, effort quotas and catch quotas to regulate access in the different sectors in a way that suits the nature of each, and gives due attention to the productivity of the resources. By way of example, *A fishery manager's guidebook* by FAO tentatively suggests:

- TURFs may be particularly suitable for the management of sedentary resources;
- effort rights may be more effective and practical than catch rights where there are no reliable estimates of biomass or where good monitoring of catches may be impractical (or where species diversity is high);
- catch rights may best facilitate the management of highly migratory and transboundary stocks where the allowable catch must be divided amongst the participating nations; and
- effort management may be more effective where a fishery uses primarily the same gear type, whereas in a fishery using many different gear types, catch rights may be preferable.

EAF requires that all the uses and users of a fishery resource be considered and reconciled, and that interactions between different fisheries within the designated geographic area be taken into account. This will mean that the systems of access rights across different fisheries or different fishery sectors within the management area should be mutually compatible and, overall, that the total effort applied should be commensurate with the productivity of the ecosystem and its component parts. While this may be a demanding and difficult task to implement, often with significant political implications, it is essential for sustainable use of ecosystems and, once in place, will greatly facilitate management of the fisheries and their operation.

Creating Incentives for EAF

EAF may be easier to implement if the rules and regulations applied under a so-called control and command (C&C) form of management are supplemented, or even replaced to the extent possible, with more appropriate incentive measures to achieve EAF. The idea of incentives is to provide signals reflecting public objectives while leaving some room for individual and collective decision-making to respond to them.

Different kinds of incentives can be developed in isolation or in combination.

— Improve the institutional framework.

— Develop collective values (education, information, training).

— Create non-market economic incentives (taxes and subsidies).

— Establish market incentives.

Incentives play indirectly through the determinants of individual/collective choices, such as the profit motive or normative values. Market or social forces can be very efficient vectors to force the global outcome of individual actions towards collectively set objectives.

Any of these instruments relies to some degree on command and control. Creating the conditions for an efficient market for property rights requires that these rights be legally set and effectively enforced. Similarly, creating a market-based incentive for environmentally-friendly production methods through product eco-labelling requires that certification standards be established and enforced. Incentives and command and control should be seen as complementary, having relative advantages or disadvantages depending on what they are supposed to achieve.

Assessing Costs and Benefits of EAF

The shift to EAF may in most, if not all, cases imply higher management costs that include acquisition of additional information, planning and consultative decision-making processes involving a broader range of stakeholders/interest groups, and additional monitoring, control and surveillance. Although higher management costs may often be out-weighed by the long-term benefits of implementing EAF, the question of who pays becomes important. The idea of the fishing industry paying some of the fishery management costs is becoming increasingly accepted and adopted. However, the fact that EAF responds to wider societal needs requires an explicit policy on how the incremental management costs of EAF should be divided between benefits derived by those dependent on fishing for food, livelihood and employment, and benefits to society at large. Where countries are given the task of managing global ecosystem goods and services, consideration may have to be given to whether incremental management costs should be carried by the international community.

In considering global ecosystem goods and services such as biodiversity or conservation of endangered species, the issue arises whether valuation should be based on national or local preferences, or take into account preferences of the citizens of other countries or the international community at large. It also needs to take note of goals expressed in international conventions. On the other hand, valuation based on what the most affluent citizens of the globe are willing to pay could result in policy prescriptions that are unfavourable to poor producers and consumers in developing countries. This has given rise to the call for establishing equivalency standards that explicitly take into account differences in wealth and the ability to provide alternative employment and income opportunities.

EAF Cost-benefit Analysis

The appropriate tools to estimate the costs and benefits of EAF management measures include bio-economic and ecological-economic modelling of various sophistication and total economic valuation methods. A useful cross-sectoral tool is integrated environmental and economic accounting. A system of integrated environmental and economic accounts (SEEA) provides a comprehensive framework to monitor and analyse the interactions between different sectors of the economy and their individual and aggregate impacts on the environment.

Other Considerations

Many of the problems facing fisheries management in an EAF context fall outside the direct control of fisheries managers. Examples of such problems include:

- eutrophication of coastal waters resulting from excess nutrients from agriculture and sewage, which cause toxic algal blooms and affect the health of seagrass and coral reef habitats (by encouraging growth of epiphytes, for example);
- sediment loads from agriculture, forestry and construction of infrastructure in catchment that degrade coastal ecosystems, particularly the critical coral reefs and seagrass habitats;

— destruction of fish habitats through foreshore development;
— introduction of exotic species through ballast water and on the hulls of ships;
— contamination of fish products through chemical pollution from agriculture and industry;
— competing use of waterways from other sectors, including aquaculture; and
— effects of climate change on distribution of stocks and sea level rise on nursery habitats.

Fisheries managers need to ensure that they are recognized as important stakeholders in the process of integrated coastal management so that they can safeguard the function of the habitats that support fisheries ecosystems from adverse effects stemming from activities in other sectors,

Threats to Implementing EAF

The need to progress towards EAF has been widely recognized and was embedded to a large extent in the Code of Conduct for Responsible Fisheries. However, there are substantial obstacles to the effective implementation of EAF, as evidenced by the difficulties of countries in implementing the requirements of the Code. Key impediments to EAF include the following:

1. The mismatch between expectations and resources (both human and financial) will need to be carefully managed. EAF has much to offer, but lack of investment in the process will certainly slow progress and might mean failure in the end. The differing timetables of the political and the management process may also mean that insufficient commitment and resources are made available. EAF is a long-term commitment with long-term benefits, which may be difficult to present convincingly to governments, which normally work in shorter cycles, and especially when EAF competes with short-term socio-economic objectives.
2. Difficulty may be foreseen in reconciling competing objectives of the multiple stakeholders. In some, perhaps many, cases the participatory process may be insufficient for finding compromises that satisfy all stakeholders. Conflicts may then require higher-level intervention to determine the relative priorities and possibly, compensation. This is

already a serious problem in many TROM fisheries, and will be exacerbated by EAF.

3. Insufficient or ineffective participation of stakeholders in the development and implementation of EAF may occur, even when competing objectives can be reconciled. This deficiency could be caused by a number of factors including:
   - an unwillingness of stakeholders to participate openly and transparently in the process or to make concessions, believing that they will fare better by non-cooperation than by cooperation;
   - inadequate and fuzzy user rights that fail to recognize long-term interests and responsibilities leading to poor stewardship;
   - a lack of access to necessary information;
   - inadequate consultation process or arrangements;
   - insufficient resources being invested to improve fisheries and their management;
   - a lack of capacity to participate effectively (e.g. knowledge, financial or other resources, geographical dispersion); and
   - hidden agendas (e.g. expectations that are not transparent to all participants, leading to distorting behaviour and mistrust).
4. The time and cost required for effective consultation with a wide range of stakeholders could be substantial but, in many cases, a good start can be made with the resources already being used for TROM.
5. Insufficient knowledge will continue to be a constraint. Biological uncertainty is recognized as a substantial problem in management of fisheries under TROM, and the combined biological and ecological uncertainty under EAF will be even greater. One manifestation of this will be an inability in some instances to identify meaningful, cost-effective indicators for important objectives. The sum of these uncertainties will require robust and precautionary approaches that could cause substantial difficulties in some cases for certain stakeholders, both social and economic. A further source of uncertainty is a widespread lack of adequate knowledge of fleet and fisher behaviour and dynamics.
6. A lack of adequate capacity for informative compilation and analysis of the available information will often add to the uncertainty. In cases

where there are or have been inadequate monitoring and data storage systems in place, the problems will be particularly acute.

7. Insufficient education and awareness will also be a problem. This will apply to all stakeholders in exercising their responsibilities, including the fishery management agencies and the public, who will need to be better educated on their roles in the process.
8. Equity issues will always be difficult to resolve in relation to responsibility for ecosystem degradation, between fisheries and other economic activities such as agriculture (including forestry), chemical industries, urban and coastal development, energy and tourism.
9. Aligning the boundaries of the ecosystems and of the jurisdictions of the management authorities (whether at regional, national or sub-national levels), as well as between jurisdictions of the different authorities responsible for competing sectors, will continue to be a challenge. Trans-boundary issues will require particular attention. As foreseen in the United Nations Fish Stock Agreement (FSA), EAF measures adopted by different countries sharing an ecosystem will need to be compatible across the whole geographical range of the ecosystem.
10. Another impediment common to both TROM and EAF, which will continue to be a threat, is illegal stakeholder behaviour: illegal fishing, lack of implementation of flag state and port state responsibilities, and misreporting. While these types of practices continue, it is difficult to see how the principles and processes outlined in these guidelines can be implemented successfully, especially on the high seas. The Compliance Agreement and the International Plan of Action on Illegal, Unreported and Unregulated fishing should play a useful role in changing this situation for the future.
11. Poverty is a major threat to EAF. While poor coastal dwellers have few other options to derive livelihoods, fishing will continue to be the occupation of last resort for growing and displaced populations, resulting in excessive fishing effort, depletion of resources and ecosystem degradation. This will often occur in desperate circumstances where the incentive to care for the ecosystem is overshadowed by daily necessities.

References

Charles, A.T., Use rights and responsible fisheries: limiting access and harvesting through rights-based management, in *A fishery manager's guidebook – Management measures and their application*, K.L. Cochrane (ed.), FAO Fisheries Technical Paper, No. 424, pp. 131–157.

Commonwealth of Australia, *National strategy for ecologically sustainable development,* Canberra, Australian Government Publishing Service, 1992.

FAO Fisheries Resources Division, I*ndicators for sustainable development of marine capture fisheries,* FAO Technical Guidelines for Responsible Fisheries, No. 8, 1999.

World Wildlife Foundation Australia, *Policy proposals and operational guidance for ecosystem-based management of marine capture fisheries,* 2002

7

Code of Conduct for Responsible Fisheries

Fishing has been a major source of food for humanity and a provider of employment and economic benefits to those engaged in this activity, from ancient times. The wealth of aquatic resources was assumed to be an unlimited gift of nature. However, with increased knowledge and the dynamic development of fisheries after the second world war, this myth has faded in face of the realization that aquatic resources, although renewable, are not infinite and need to be properly managed, if their contribution to the nutritional, economic and social well-being of the growing world's population is to be sustained.

The widespread introduction in the mid-seventies of exclusive economic zones (EEZs) and the adoption in 1982, after long deliberations, of the United Nations Convention on the Law of the Sea provided a new framework for the better management of marine resources. The new legal regime of the ocean gave coastal States rights and responsibilities for the management and use of fishery resources within their EEZs which embrace some 90 percent of the world's marine fisheries. Such extended national jurisdiction was a necessary but insufficient step toward the efficient management and sustainable development of fisheries. Many coastal States continued to face serious challenges as, lacking experience and financial and physical resources, they sought to extract greater benefits from the fisheries within their EEZs.

In recent years, world fisheries have become a market-driven, dynamically developing sector of the food industry and coastal States have striven to take advantage of their new opportunities by investing in modern fishing fleets and processing factories in response to growing international demand for fish and fishery products. By the late 1980s it became clear, however, that fisheries resources could no longer sustain such rapid and often uncontrolled exploitation and development, and that new approaches to fisheries management embracing conservation and environmental considerations were urgently needed. The situation was aggravated by the realization that unregulated fisheries on the high seas, in some cases involving straddling and highly migratory fish species, which occur within and outside EEZs, were becoming a matter of increasing concern.

The Committee on Fisheries (COFI) at its Nineteenth Session in March 1991 called for the development of new concepts which would lead to responsible, sustained fisheries. Subsequently, the International Conference on Responsible Fishing, held in 1992 in Cancûn (Mexico) further requested FAO to prepare an international Code of Conduct to address these concerns. The outcome of this Conference, particularly the Declaration of Cancûn, was an important contribution to the 1992 United Nations Conference on Environment and Development (UNCED), in particular its Agenda 21. Subsequently, the United Nations Conference on Straddling Fish Stocks and Highly Migratory Fish Stocks was convened, to which FAO provided important technical back-up. In November 1993, the Agreement to Promote Compliance with International Conservation and Management Measures by Fishing Vessels on the High Seas was adopted at the Twenty-seventh Session of the FAO Conference.

Noting these and other important developments in world fisheries, the FAO Governing Bodies recommended the formulation of a global Code of Conduct for Responsible Fisheries which would be consistent with these instruments and, in a non-mandatory manner, establish principles and standards applicable to the conservation, management and development of all fisheries. The Code, which was unanimously adopted on 31 October 1995 by the FAO Conference, provides a necessary framework for national and international efforts to ensure sustainable exploitation of aquatic living resources in harmony with the environment.

FAO, in accordance with its mandate, is fully committed to assisting Member States, particularly developing countries, in the efficient implementation of the Code of Conduct for Responsible Fisheries and will report to the United Nations community on the progress achieved and further action required.

Fisheries, including aquaculture, provide a vital source of food, employment, recreation, trade and economic well being for people throughout the world, both for present and future generations and should therefore be conducted in a responsible manner. This Code sets out principles and international standards of behaviour for responsible practices with a view to ensuring the effective conservation, management and development of living aquatic resources, with due respect for the ecosystem and biodiversity. The Code recognises the nutritional, economic, social, environmental and cultural importance of fisheries, and the interests of all those concerned with the fishery sector. The Code takes into account the biological characteristics of the resources and their environment and the interests of consumers and other users. States and all those involved in fisheries are encouraged to apply the Code and give effect to it.

Article 1 : Nature and scope of the Code

1.1 This Code is voluntary. However, certain parts of it are based on relevant rules of international law, including those reflected in the United Nations Convention on the Law of the Sea of 10 December 1982. The Code also contains provisions that may be or have already been given binding effect by means of other obligatory legal instruments amongst the Parties, such as the Agreement to Promote Compliance with International Conservation and Management Measures by Fishing Vessels on the High Seas, 1993, which, according to FAO Conference resolution 15/93, paragraph 3, forms an integral part of the Code.

1.2 The Code is global in scope, and is directed toward members and non-members of FAO, fishing entities, subregional, regional and global organisations, whether governmental or non-governmental, and all persons concerned with the conservation of fishery resources and management and development of fisheries, such as fishers, those engaged in processing and marketing of fish and fishery products and other users of the aquatic environment in relation to fisheries.

1.3 The Code provides principles and standards applicable to the conservation, management and development of all fisheries. It also covers the capture, processing and trade of fish and fishery products, fishing operations, aquaculture, fisheries research and the integration of fisheries into coastal area management.

1.4 In this Code, the reference to States includes the European Community in matters within its competence, and the term fisheries applies equally to capture fisheries and aquaculture.

Article 2 : Objectives of the Code

The objectives of the Code are to:

— establish principles, in accordance with the relevant rules of international law, for responsible fishing and fisheries activities, taking into account all their relevant biological, technological, economic, social, environmental and commercial aspects;

— establish principles and criteria for the elaboration and implementation of national policies for responsible conservation of fisheries resources and fisheries management and development;

— serve as an instrument of reference to help States to establish or to improve the legal and institutional framework required for the exercise of responsible fisheries and in the formulation and implementation of appropriate measures;

— provide guidance which may be used where appropriate in the formulation and implementation of international agreements and other legal instruments, both binding and voluntary;

— facilitate and promote technical, financial and other cooperation in conservation of fisheries resources and fisheries management and development;

— promote the contribution of fisheries to food security and food quality, giving priority to the nutritional needs of local communities;

— promote protection of living aquatic resources and their environments and coastal areas;

— promote the trade of fish and fishery products in conformity with relevant international rules and avoid the use of measures that constitute hidden barriers to such trade;

— promote research on fisheries as well as on associated ecosystems and relevant environmental factors; and

— provide standards of conduct for all persons involved in the fisheries sector.

Article 3 : Relationship with other international instruments

3.1 The Code is to be interpreted and applied in conformity with the relevant rules of international law, as reflected in the United Nations Convention on the Law of the Sea, 1982. Nothing in this Code prejudices the rights, jurisdiction and duties of States under international law as reflected in the Convention.

3.2 The Code is also to be interpreted and applied:

— in a manner consistent with the relevant provisions of the Agreement for the Implementation of the Provisions of the United Nations Convention on the Law of the Sea of 10 December 1982 Relating to the Conservation and Management of Straddling Fish Stocks and Highly Migratory Fish Stocks;

— in accordance with other applicable rules of international law, including the respective obligations of States pursuant to international agreements to which they are party; and

— in the light of the 1992 Declaration of Cancun, the 1992 Rio Declaration on Environment and Development, and Agenda 21 adopted by the United Nations Conference on Environment and Development (UNCED), in particular Chapter 17 of Agenda 21, and other relevant declarations and international instruments.

Article 4 : Implementation monitoring and updating

4.1 All members and non-members of FAO, fishing entities and relevant subregional, regional and global organisations, whether governmental or non-governmental, and all persons concerned with the conservation, management and utilisation of fisheries resources and trade in fish and fishery products should collaborate in the fulfilment and implementation of the objectives and principles contained in this Code.

4.2 FAO, in accordance with its role within the United Nations system, will monitor the application and implementation of the Code and its effects on fisheries and the Secretariat will report accordingly to the Committee on Fisheries (COFI). All States, whether members or non-members of FAO, as well as relevant international organisations, whether governmental or non-governmental should actively cooperate with FAO in this work.

4.3 FAO, through its competent bodies, may revise the Code, taking into account developments in fisheries as well as reports to COFI on the implementation of the Code.

4.4 States and international organisations, whether governmental or non-governmental, should promote the understanding of the Code among those involved in fisheries, including, where practicable, by the introduction of schemes which would promote voluntary acceptance of the Code and its effective application.

Article 5 : Special requirements of developing countries

5.1 The capacity of developing countries to implement the recommendations of this Code should be duly taken into account.

5.2 In order to achieve the objectives of this Code and to support its effective implementation, countries, relevant international organisations, whether governmental or non-governmental, and financial institutions should give full recognition to the special circumstances and requirements of developing countries, including in particular the least-developed among them, and small island developing countries. States, relevant intergovernmental and non-governmental organisations and financial institutions should work for the adoption of measures to address the needs of developing countries, especially in the areas of financial and technical assistance, technology transfer, training and scientific cooperation and in enhancing their ability to develop their own fisheries as well as to participate in high seas fisheries, including access to such fisheries.

Article 6 : General principles

6.1 States and users of living aquatic resources should conserve aquatic ecosystems. The right to fish carries with it the obligation to do so in a responsible manner so as to ensure effective conservation and management of the living aquatic resources.

6.2 Fisheries management should promote the maintenance of the quality, diversity and availability of fishery resources in sufficient quantities for present and future generations in the context of food security, poverty alleviation and sustainable development. Management measures should not only ensure the conservation of target species but also of species belonging to the same ecosystem or associated with or dependent upon the target species.

6.3 States should prevent overfishing and excess fishing capacity and should implement management measures to ensure that fishing effort is commensurate with the productive capacity of the fishery resources and their sustainable utilisation. States should take measures to rehabilitate populations as far as possible and when appropriate.

6.4 Conservation and management decisions for fisheries should be based on the best scientific evidence available, also taking into account traditional knowledge of the resources and their habitat, as well as relevant environmental, economic and social factors. States should assign priority to undertake research and data collection in order to improve scientific and technical knowledge of fisheries including their interaction with the ecosystem. In recognising the transboundary nature of many aquatic ecosystems, States should encourage bilateral and multilateral cooperation in research, as appropriate.

6.5 States and subregional and regional fisheries management organisations should apply a precautionary approach widely to conservation, management and exploitation of living aquatic resources in order to protect them and preserve the aquatic environment, taking account of the best scientific evidence available. The absence of adequate scientific information should not be used as a reason for postponing or failing to take measures to conserve target species, associated or dependent species and non-target species and their environment.

6.6 Selective and environmentally safe fishing gear and practices should be further developed and applied, to the extent practicable, in order to maintain biodiversity and to conserve the population structure and aquatic ecosystems and protect fish quality. Where proper selective and environmentally safe fishing gear and practices exist, they should be recognised and accorded a priority in establishing conservation and management measures for fisheries. States and users of aquatic ecosystems should minimise waste, catch of non-target species, both fish and non-fish species, and impacts on associated or dependent species.

6.7 The harvesting, handling, processing and distribution of fish and fishery products should be carried out in a manner which will maintain the nutritional value, quality and safety of the products, reduce waste and minimise negative impacts on the environment.

6.8 All critical fisheries habitats in marine and fresh water ecosystems, such as wetlands, mangroves, reefs, lagoons, nursery and spawning areas, should be protected and rehabilitated as far as possible and where necessary. Particular effort should be made to protect such habitats from destruction, degradation, pollution and other significant impacts resulting from human activities that threaten the health and viability of the fishery resources.

6.9 States should ensure that their fisheries interests, including the need for conservation of the resources, are taken into account in the multiple uses of the coastal zone and are integrated into coastal area management, planning and development.

6.10 Within their respective competences and in accordance with international law, including within the framework of subregional or regional fisheries conservation and management organisations or arrangements, States should ensure compliance with and enforcement of conservation and management measures and establish effective mechanisms, as appropriate, to monitor and control the activities of fishing vessels and fishing support vessels.

6.11 States authorising fishing and fishing support vessels to fly their flags should exercise effective control over those vessels so as to ensure the proper application of this Code. They should ensure that the activities of such vessels do not undermine the effectiveness of conservation and management measures taken in accordance with international law and adopted at the national, subregional, regional or global levels. States should also ensure that vessels flying their flags fulfil their obligations concerning the collection and provision of data relating to their fishing activities.

6.12 States should, within their respective competences and in accordance with international law, cooperate at subregional, regional and global levels through fisheries management organisations, other international agreements or other arrangements to promote conservation and management, ensure responsible fishing and ensure effective conservation and protection of living aquatic resources throughout their range of distribution, taking into account the need for compatible measures in areas within and beyond national jurisdiction.

6.13 States should, to the extent permitted by national laws and regulations, ensure that decision making processes are transparent and achieve timely solutions to urgent matters. States, in accordance with appropriate procedures, should facilitate consultation and the effective participation of industry, fishworkers, environmental and other interested organisations in decision making with respect to the development of laws and policies related to fisheries management, development, international lending and aid.

6.14 International trade in fish and fishery products should be conducted in accordance with the principles, rights and obligations established in the World Trade Organisation (WTO) Agreement and other relevant international agreements. States should ensure that their policies, programmes and practices related to trade in fish and fishery products do not result in obstacles to this trade, environmental degradation or negative social, including nutritional, impacts.

6.15 States should cooperate in order to prevent disputes. All disputes relating to fishing activities and practices should be resolved in a timely, peaceful and cooperative manner, in accordance with applicable international agreements or as may otherwise be agreed between the parties. Pending settlement of a dispute, the States concerned should make every effort to enter into provisional arrangements of a practical nature which should be without prejudice to the final outcome of any dispute settlement procedure.

6.16 States, recognising the paramount importance to fishers and fishfarmers of understanding the conservation and management of the fishery resources on which they depend, should promote awareness of responsible fisheries through education and training. They should ensure that fishers and fishfarmers are involved in the policy formulation and implementation process, also with a view to facilitating the implementation of the Code.

6.17 States should ensure that fishing facilities and equipment as well as all fisheries activities allow for safe, healthy and fair working and living conditions and meet internationally agreed standards adopted by relevant international organisations.

6.18 Recognising the important contributions of artisanal and small-scale fisheries to employment, income and food security, States should appropriately protect the rights of fishers and fishworkers, particularly those engaged in subsistence, small-scale and artisanal fisheries, to a secure and just livelihood, as well as preferential access, where appropriate, to traditional fishing grounds and resources in the waters under their national jurisdiction.

6.19 States should consider aquaculture, including culture-based fisheries, as a means to promote diversification of income and diet. In so doing, States should ensure that resources are used responsibly and adverse impacts on the environment and on local communities are minimised.

Article 7 : Fisheries management

7.1 General

7.1.1 States and all those engaged in fisheries management should, through an appropriate policy, legal and institutional framework, adopt measures for the long-term conservation and sustainable use of fisheries resources. Conservation and management measures, whether at local, national, subregional or regional levels, should be based on the best scientific evidence available and be designed to ensure the long-term sustainability of fishery resources at levels which promote the objective of their optimum utilisation and maintain their availability for present and future generations; short term considerations should not compromise these objectives.

7.1.2 Within areas under national jurisdiction, States should seek to identify relevant domestic parties having a legitimate interest in the use and management of fisheries resources and establish arrangements for consulting them to gain their collaboration in achieving responsible fisheries.

7.1.3 For transboundary fish stocks, straddling fish stocks, highly migratory fish stocks and high seas fish stocks, where these are exploited by two or more States, the States concerned, including the relevant coastal States in the case of straddling and highly migratory stocks, should cooperate to ensure effective conservation and management of the resources. This should be achieved, where appropriate, through the establishment of a bilateral, subregional or regional fisheries organisation or arrangement.

7.1.4 A subregional or regional fisheries management organisation or arrangement should include representatives of States in whose jurisdictions the resources occur, as well as representatives from States which have a real interest in the fisheries on the resources outside national jurisdictions. Where a subregional or regional fisheries management organisation or arrangement exists and has the competence to establish conservation and management measures, those States should cooperate by becoming a member of such organisation or a participant in such arrangement, and actively participate in its work.

7.1.5 A State which is not a member of a subregional or regional fisheries management organisation or is not a participant in a subregional or regional fisheries management arrangement should nevertheless cooperate, in accordance with relevant international agreements and international law, in the conservation and management of the relevant fisheries resources by giving effect to any conservation and management measures adopted by such organisation or arrangement.

7.1.6 Representatives from relevant organisations, both governmental and non-governmental, concerned with fisheries should be afforded the opportunity to take part in meetings of subregional and regional fisheries management organisations and arrangements as observers or otherwise, as appropriate, in accordance with the procedures of the organisation or arrangement concerned. Such representatives should be given timely access to the records and reports of such meetings, subject to the procedural rules on access to them.

7.1.7 States should establish, within their respective competences and capacities, effective mechanisms for fisheries monitoring, surveillance, control and enforcement to ensure compliance with their conservation and management measures, as well as those adopted by subregional or regional organisations or arrangements.

7.1.8 States should take measures to prevent or eliminate excess fishing capacity and should ensure that levels of fishing effort are commensurate with the sustainable use of fishery resources as a means of ensuring the effectiveness of conservation and management measures.

7.1.9 States and subregional or regional fisheries management organisations and arrangements should ensure transparency in the mechanisms for fisheries management and in the related decision-making process.

7.1.10 States and subregional or regional fisheries management organisations and arrangements should give due publicity to conservation and management measures and ensure that laws, regulations and other legal rules governing their implementation are effectively disseminated. The bases and purposes of such measures should be explained to users of the resource in order to facilitate their application and thus gain increased support in the implementation of such measures.

7.2 Management Objectives

7.2.1 Recognising that long-term sustainable use of fisheries resources is the overriding objective of conservation and management, States and subregional or regional fisheries management organisations and arrangements should, inter alia, adopt appropriate measures, based on the best scientific evidence available, which are designed to maintain or restore stocks at levels capable of producing maximum sustainable yield, as qualified by relevant environmental and economic factors, including the special requirements of developing countries.

7.2.2 Such measures should provide inter alia that:

— excess fishing capacity is avoided and exploitation of the stocks remains economically viable;

— the economic conditions under which fishing industries operate promote responsible fisheries;

— the interests of fishers, including those engaged in subsistence, small-scale and artisanal fisheries, are taken into account;

— biodiversity of aquatic habitats and ecosystems is conserved and endangered species are protected;

— depleted stocks are allowed to recover or, where appropriate, are actively restored;

— adverse environmental impacts on the resources from human activities are assessed and, where appropriate, corrected; and

— pollution, waste, discards, catch by lost or abandoned gear, catch of non-target species, both fish and non- fish species, and impacts on associated or dependent species are minimised, through measures including, to the extent practicable, the development and use of selective, environmentally safe and cost-effective fishing gear and techniques.

7.2.3 States should assess the impacts of environmental factors on target stocks and species belonging to the same ecosystem or associated with or dependent upon the target stocks, and assess the relationship among the populations in the ecosystem.

7.3 Management Framework and Procedures

7.3.1 To be effective, fisheries management should be concerned with the whole stock unit over its entire area of distribution and take into account previously agreed management measures established and applied in the same region, all removals and the biological unity and other biological characteristics of the stock. The best scientific evidence available should be used to determine, inter alia, the area of distribution of the resource and the area through which it migrates during its life cycle.

7.3.2 In order to conserve and manage transboundary fish stocks, straddling fish stocks, highly migratory fish stocks and high seas fish stocks throughout their range, conservation and management measures established for such stocks in accordance with the respective competences of relevant States or, where appropriate, through subregional and regional fisheries management organisations and arrangements, should be compatible. Compatibility should be achieved in a manner consistent with the rights, competences and interests of the States concerned.

7.3.3 Long-term management objectives should be translated into management actions, formulated as a fishery management plan or other management framework.

7.3.4 States and, where appropriate, subregional or regional fisheries management organisations and arrangements should foster and promote international cooperation and coordination in all matters related to fisheries, including information gathering and exchange, fisheries research, management and development.

7.3.5 States seeking to take any action through a non-fishery organisation which may affect the conservation and management measures taken by a competent subregional or regional fisheries management organisation or arrangement should consult with the latter, in advance to the extent practicable, and take its views into account.

7.4 Data Gathering and Management Advice

7.4.1 When considering the adoption of conservation and management measures, the best scientific evidence available should be taken into account in order to evaluate the current state of the fishery resources and the possible impact of the proposed measures on the resources.

7.4.2 Research in support of fishery conservation and management should be promoted, including research on the resources and on the effects of climatic, environmental and socio-economic factors. The results of such research should be disseminated to interested parties.

7.4.3 Studies should be promoted which provide an understanding of the costs, benefits and effects of alternative management options designed to rationalise fishing, in particular, options relating to excess fishing capacity and excessive levels of fishing effort.

7.4.4 States should ensure that timely, complete and reliable statistics on catch and fishing effort are collected and maintained in accordance with applicable international standards and practices and in sufficient detail to allow sound statistical analysis. Such data should be updated regularly and verified through an appropriate system. States should compile and disseminate such data in a manner consistent with any applicable confidentiality requirements.

7.4.5 In order to ensure sustainable management of fisheries and to enable social and economic objectives to be achieved, sufficient knowledge of social, economic and institutional factors should be developed through data gathering, analysis and research.

7.4.6 States should compile fishery-related and other supporting scientific data relating to fish stocks covered by subregional or regional fisheries management organisations or arrangements in an internationally agreed format and provide them in a timely manner to the organisation or arrangement. In cases of stocks which occur in the jurisdiction of more than one State and for which there is no such organisation or arrangement, the States concerned should agree on a mechanism for cooperation to compile and exchange such data.

7.4.7 Subregional or regional fisheries management organisations or arrangements should compile data and make them available, in a manner consistent with any applicable confidentiality requirements, in a timely manner and in an agreed format to all members of these organisations and other interested parties in accordance with agreed procedures.

7.5 Precautionary Approach

7.5.1 States should apply the precautionary approach widely to conservation, management and exploitation of living aquatic resources in order to protect them and preserve the aquatic environment. The absence of adequate scientific information should not be used as a reason for postponing or failing to take conservation and management measures.

7.5.2 In implementing the precautionary approach, States should take into account, inter alia, uncertainties relating to the sise and productivity of the stocks, reference points, stock condition in relation to such reference points, levels and distribution of fishing mortality and the impact of fishing activities, including discards, on non-target and associated or dependent species, as well as environmental and socio-economic conditions.

7.5.3 States and subregional or regional fisheries management organisations and arrangements should, on the basis of the best scientific evidence available, inter alia, determine:

— stock specific target reference points, and, at the same time, the action to be taken if they are exceeded; and

— stock-specific limit reference points, and, at the same time, the action to be taken if they are exceeded; when a limit reference point is approached, measures should be taken to ensure that it will not be exceeded.

7.5.4 In the case of new or exploratory fisheries, States should adopt as soon as possible cautious conservation and management measures, including, inter alia, catch limits and effort limits. Such measures should remain in force until there are sufficient data to allow assessment of the impact of the fisheries on the long-term sustainability of the stocks, whereupon conservation and management measures based on that assessment should be implemented. The latter measures should, if appropriate, allow for the gradual development of the fisheries.

7.5.5 If a natural phenomenon has a significant adverse impact on the status of living aquatic resources, States should adopt conservation and management measures on an emergency basis to ensure that fishing activity does not exacerbate such adverse impact. States should also adopt such measures on an emergency basis where fishing activity presents a serious threat to the sustainability of such resources. Measures taken on an emergency basis should be temporary and should be based on the best scientific evidence available.

7.6 Management Measures

7.6.1 States should ensure that the level of fishing permitted is commensurate with the state of fisheries resources.

7.6.2 States should adopt measures to ensure that no vessel be allowed to fish unless so authorised, in a manner consistent with international law for the high seas or in conformity with national legislation within areas of national jurisdiction.

7.6.3 Where excess fishing capacity exists, mechanisms should be established to reduce capacity to levels commensurate with the sustainable use of fisheries resources so as to ensure that fishers operate under economic conditions that promote responsible fisheries. Such mechanisms should include monitoring the capacity of fishing fleets.

7.6.4 The performance of all existing fishing gear, methods and practices should be examined and measures taken to ensure that fishing gear, methods and practices which are not consistent with responsible fishing are phased out and replaced with more acceptable alternatives. In this process, particular attention should be given to the impact of such measures on fishing communities, including their ability to exploit the resource.

7.6.5 States and fisheries management organisations and arrangements should regulate fishing in such a way as to avoid the risk of conflict among fishers using different vessels, gear and fishing methods.

7.6.6 When deciding on the use, conservation and management of fisheries resources, due recognition should be given, as appropriate, in accordance with national laws and regulations, to the traditional practices, needs and interests of indigenous people and local fishing communities which are highly dependent on fishery resources for their livelihood.

7.6.7 In the evaluation of alternative conservation and management measures, their cost-effectiveness and social impact should be considered.

7.6.8 The efficacy of conservation and management measures and their possible interactions should be kept under continuous review. Such measures should, as appropriate, be revised or abolished in the light of new information.

7.6.9 States should take appropriate measures to minimise waste, discards, catch by lost or abandoned gear, catch of non-target species, both fish and non-fish species, and negative impacts on associated or dependent species, in particular endangered species. Where appropriate, such measures may include technical measures related to fish size, mesh size or gear, discards, closed seasons and areas and zones reserved for selected fisheries, particularly artisanal fisheries. Such measures should be applied, where appropriate, to protect juveniles and spawners. States and subregional or regional fisheries management organisations and arrangements should promote, to the extent practicable, the development and use of selective, environmentally safe and cost effective gear and techniques.

7.6.10 States and subregional and regional fisheries management organisations and arrangements, in the framework of their respective competences, should introduce measures for depleted resources and those resources threatened with depletion that facilitate the sustained recovery of such stocks. They should make every effort to ensure that resources and habitats critical to the well-being of such resources which have been adversely affected by fishing or other human activities are restored.

7.7 Implementation

7.7.1 States should ensure that an effective legal and administrative framework at the local and national level, as appropriate, is established for fisheries resource conservation and fisheries management.

7.7.2 States should ensure that laws and regulations provide for sanctions applicable in respect of violations which are adequate in severity to be effective, including sanctions which allow for the refusal, withdrawal or suspension of authorisations to fish in the event of non-compliance with conservation and management measures in force.

7.7.3 States, in conformity with their national laws, should implement effective fisheries monitoring, control, surveillance and law enforcement measures including, where appropriate, observer programmes, inspection schemes and vessel monitoring systems. Such measures should be promoted and, where appropriate, implemented by subregional or regional fisheries management organisations and arrangements in accordance with procedures agreed by such organisations or arrangements.

7.7.4 States and subregional or regional fisheries management organisations and arrangements, as appropriate, should agree on the means by which the activities of such organisations and arrangements will be financed, bearing in mind, inter alia, the relative benefits derived from the fishery and the differing capacities of countries to provide financial and other contributions. Where appropriate, and when possible, such organisations and arrangements should aim to recover the costs of fisheries conservation, management and research.

7.7.5 States which are members of or participants in subregional or regional fisheries management organisations or arrangements should implement internationally agreed measures adopted in the framework of such organisations or arrangements and consistent with international law to deter the activities of vessels flying the flag of non-members or non-participants which engage in activities which undermine the effectiveness of conservation and management measures established by such organisations or arrangements.

7.8 Financial Institutions

7.8.1 Without prejudice to relevant international agreements, States should encourage banks and financial institutions not to require, as a condition of a loan or mortgage, fishing vessels or fishing support vessels to be flagged in a jurisdiction other than that of the State of beneficial ownership where such a requirement would have the effect of increasing the likelihood of non-compliance with international conservation and management measures.

Article 8 : Fishing operations

8.1 Duties of all States

8.1.1 States should ensure that only fishing operations allowed by them are conducted within waters under their jurisdiction and that these operations are carried out in a responsible manner.

8.1.2 States should maintain a record, updated at regular intervals, on all authorisations to fish issued by them.

8.1.3 States should maintain, in accordance with recognised international standards and practices, statistical data, updated at regular intervals, on all fishing operations allowed by them.

8.1.4 States should, in accordance with international law, within the framework of subregional or regional fisheries management organisations or arrangements, cooperate to establish systems for monitoring, control, surveillance and enforcement of applicable measures with respect to fishing operations and related activities in waters outside their national jurisdiction.

8.1.5 States should ensure that health and safety standards are adopted for everyone employed in fishing operations. Such standards should be not less than the minimum requirements of relevant international agreements on conditions of work and service.

8.1.6 States should make arrangements individually, together with other States or with the appropriate international organisation to integrate fishing operations into maritime search and rescue systems.

8.1.7 States should enhance through education and training programmes the education and skills of fishers and, where appropriate, their professional qualifications. Such programmes should take into account agreed international standards and guidelines.

8.1.8 States should, as appropriate, maintain records of fishers which should, whenever possible, contain information on their service and qualifications, including certificates of competency, in accordance with their national laws.

8.1.9 States should ensure that measures applicable in respect of masters and other officers charged with an offence relating to the operation of fishing vessels should include provisions which may permit, inter alia, refusal, withdrawal or suspension of authorisations to serve as masters or officers of a fishing vessel.

8.1.10 States, with the assistance of relevant international organisations, should endeavour to ensure through education and training that all those engaged in fishing operations be given information on the most important provisions of this Code, as well as provisions of relevant international conventions and applicable environmental and other standards that are essential to ensure responsible fishing operations.

8.2 Flag State Duties

8.2.1 Flag States should maintain records of fishing vessels entitled to fly their flag and authorised to be used for fishing and should indicate in such records details of the vessels, their ownership and authorisation to fish.

8.2.2 Flag States should ensure that no fishing vessels entitled to fly their flag fish on the high seas or in waters under the jurisdiction of other States unless such vessels have been issued with a Certificate of Registry and have been authorised to fish by the competent authorities. Such vessels should carry on board the Certificate of Registry and their authorisation to fish.

8.2.3 Fishing vessels authorised to fish on the high seas or in waters under the jurisdiction of a State other than the flag State, should be marked in accordance with uniform and internationally recognisable vessel marking systems such as the FAO Standard Specifications and Guidelines for Marking and Identification of Fishing Vessels.

8.2.4 Fishing gear should be marked in accordance with national legislation in order that the owner of the gear can be identified. Gear marking requirements should take into account uniform and internationally recognisable gear marking systems.

8.2.5 Flag States should ensure compliance with appropriate safety requirements for fishing vessels and fishers in accordance with international conventions, internationally agreed codes of practice and voluntary guidelines. States should adopt appropriate safety requirements for all small vessels not covered by such international conventions, codes of practice or voluntary guidelines.

8.2.6 States not party to the Agreement to Promote Compliance with International Conservation and Management Measures by Vessels Fishing in the High Seas should be encouraged to accept the Agreement and to adopt laws and regulations consistent with the provisions of the Agreement.

8.2.7 Flag States should take enforcement measures in respect of fishing vessels entitled to fly their flag which have been found by them to have contravened applicable conservation and management measures, including, where appropriate, making the contravention of such measures an offence under national legislation. Sanctions applicable in respect of violations should be adequate in severity to be effective in securing compliance and to discourage violations wherever they occur and should deprive offenders of the benefits accruing from their illegal activities. Such sanctions may, for serious violations, include provisions for the refusal, withdrawal or suspension of the authorisation to fish.

8.2.8 Flag States should promote access to insurance coverage by owners and charterers of fishing vessels. Owners or charterers of fishing vessels should carry sufficient insurance cover to protect the crew of such vessels and their interests, to indemnify third parties against loss or damage and to protect their own interests.

8.2.9 Flag States should ensure that crew members are entitled to repatriation, taking account of the principles laid down in the "Repatriation of Seafarers Convention (Revised), 1987, (No.166)".

8.2.10 In the event of an accident to a fishing vessel or persons on board a fishing vessel, the flag State of the fishing vessel concerned should provide details of the accident to the State of any foreign national on board the vessel involved in the accident. Such information should also, where practicable, be communicated to the International Maritime Organisation.

8.3 Port State Duties

8.3.1 Port States should take, through procedures established in their national legislation, in accordance with international law, including applicable international agreements or arrangements, such measures as are necessary to achieve and to assist other States in achieving the objectives of this Code, and should make known to other States details of regulations and measures they have established for this purpose. When taking such measures a port State should not discriminate in form or in fact against the vessels of any other State.

8.3.2 Port States should provide such assistance to flag States as is appropriate, in accordance with the national laws of the port State and international law, when a fishing vessel is voluntarily in a port or at an offshore terminal of the port State and the flag State of the vessel requests the port State for assistance in respect of non- compliance with subregional, regional or global conservation and management measures or with internationally agreed minimum standards for the prevention of pollution and for safety, health and conditions of work on board fishing vessels.

8.4 Fishing Activities

8.4.1 States should ensure that fishing is conducted with due regard to the safety of human life and the International Maritime Organisation International Regulations for Preventing Collisions at Sea, as well as International Maritime Organisation requirements relating to the organisation of marine traffic, protection of the marine environment and the prevention of damage to or loss of fishing gear.

8.4.2 States should prohibit dynamiting, poisoning and other comparable destructive fishing practices.

8.4.3 States should make every effort to ensure that documentation with regard to fishing operations, retained catch of fish and non-fish species and, as regards discards, the information required for stock assessment as decided by relevant management bodies, is collected and forwarded systematically to those bodies. States should, as far as possible, establish programmes, such as observer and inspection schemes, in order to promote compliance with applicable measures.

8.4.4 States should promote the adoption of appropriate technology, taking into account economic conditions, for the best use and care of the retained catch.

8.4.5 States, with relevant groups from industry, should encourage the development and implementation of technologies and operational methods that reduce discards. The use of fishing gear and practices that lead to the discarding of catch should be discouraged and the use of fishing gear and practices that increase survival rates of escaping fish should be promoted.

8.4.6 States should cooperate to develop and apply technologies, materials and operational methods that minimise the loss of fishing gear and the ghost fishing effects of lost or abandoned fishing gear.

8.4.7 States should ensure that assessments of the implications of habitat disturbance are carried out prior to the introduction on a commercial scale of new fishing gear, methods and operations to an area.

8.4.8 Research on the environmental and social impacts of fishing gear and, in particular, on the impact of such gear on biodiversity and coastal fishing communities should be promoted.

8.5 Fishing Gear Selectivity

8.5.1 States should require that fishing gear, methods and practices, to the extent practicable, are sufficiently selective so as to minimise waste, discards, catch of non-target species, both fish and non-fish species, and impacts on associated or dependent species and that the intent of related regulations is not circumvented by technical devices. In this regard, fishers should cooperate in the development of selective fishing gear and methods. States should ensure that information on new developments and requirements is made available to all fishers.

8.5.2 In order to improve selectivity, States should, when drawing up their laws and regulations, take into account the range of selective fishing gear, methods and strategies available to the industry.

8.5.3 States and relevant institutions should collaborate in developing standard methodologies for research into fishing gear selectivity, fishing methods and strategies.

8.5.4 International cooperation should be encouraged with respect to research programmes for fishing gear selectivity, and fishing methods and strategies, dissemination of the results of such research programmes and the transfer of technology.

8.6 Energy Optimisation

8.6.1 States should promote the development of appropriate standards and guidelines which would lead to the more efficient use of energy in harvesting and post-harvest activities within the fisheries sector.

8.6.2 States should promote the development and transfer of technology in relation to energy optimisation within the fisheries sector and, in particular, encourage owners, charterers and managers of fishing vessels to fit energy optimisation devices to their vessels.

8.7 Protection of the Aquatic Environment

8.7.1 States should introduce and enforce laws and regulations based on the International Convention for the Prevention of Pollution from Ships, 1973, as modified by the Protocol of 1978 relating thereto.

8.7.2 Owners, charterers and managers of fishing vessels should ensure that their vessels are fitted with appropriate equipment as required by MARPOL 73/78 and should consider fitting a shipboard compactor or incinerator to relevant classes of vessels in order to treat garbage and other shipboard wastes generated during the vessel's normal service.

8.7.3 Owners, charterers and managers of fishing vessels should minimise the taking aboard of potential garbage through proper provisioning practices.

8.7.4 The crew of fishing vessels should be conversant with proper shipboard procedures in order to ensure discharges do not exceed the levels set by MARPOL 73/78. Such procedures should, as a minimum, include the disposal of oily waste and the handling and storage of shipboard garbage.

8.8 Protection of the Atmosphere

8.8.1 States should adopt relevant standards and guidelines which would include provisions for the reduction of dangerous substances in exhaust gas emissions.

8.8.2 Owners, charterers and managers of fishing vessels should ensure that their vessels are fitted with equipment to reduce emissions of ozone depleting substances. The responsible crew members of fishing vessels should be conversant with the proper running and maintenance of machinery on board.

8.8.3 Competent authorities should make provision for the phasing out of the use of chlorofluorocarbons (CFCs) and transitional substances such as hydrochloro-fluorocarbons (HCFCs) in the refrigeration systems of fishing vessels and should ensure that the shipbuilding industry and those engaged in the fishing industry are informed of and comply with such provisions.

8.8.4 Owners or managers of fishing vessels should take appropriate action to refit existing vessels with alternative refrigerants to CFCs and HCFCs and alternatives to Halons in fire fighting installations. Such alternatives should be used in specifications for all new fishing vessels.

8.8.5 States and owners, charterers and managers of fishing vessels as well as fishers should follow international guidelines for the disposal of CFCs, HCFCs and Halons.

8.9 Harbours and Landing Places for Fishing Vessels

8.9.1 States should take into account, inter alia, the following in the design and construction of harbours and landing places:

- safe havens for fishing vessels and adequate servicing facilities for vessels, vendors and buyers are provided;
- adequate freshwater supplies and sanitation arrangements should be provided;
- waste disposal systems should be introduced, including for the disposal of oil, oily water and fishing gear;
- pollution from fisheries activities and external sources should be minimised; and
- arrangements should be made to combat the effects of erosion and siltation.

8.9.2 States should establish an institutional framework for the selection or improvement of sites for harbours for fishing vessels which allows for consultation among the authorities responsible for coastal area management.

8.10 Abandonment of Structures and other Materials

8.10.1 States should ensure that the standards and guidelines for the removal of redundant offshore structures issued by the International Maritime Organisation are followed. States should also ensure that the competent fisheries authorities are consulted prior to decisions being made on the abandonment of structures and other materials by the relevant authorities.

8.11 Artificial Reefs and Fish Aggregation Devices

8.11.1 States, where appropriate, should develop policies for increasing stock populations and enhancing fishing opportunities through the use of artificial structures, placed with due regard to the safety of navigation, on or above the seabed or at the surface. Research into the use of such structures, including the impacts on living marine resources and the environment, should be promoted.

8.11.2 States should ensure that, when selecting the materials to be used in the creation of artificial reefs as well as when selecting the geographical location of such artificial reefs, the provisions of relevant international conventions concerning the environment and safety of navigation are observed.

8.11.3 States should, within the framework of coastal area management plans, establish management systems for artificial reefs and fish aggregation devices. Such management systems should require approval for the construction and deployment of such reefs and devices and should take into account the interests of fishers, including artisanal and subsistence fishers.

8.11.4 States should ensure that the authorities responsible for maintaining cartographic records and charts for the purpose of navigation, as well as relevant environmental authorities, are informed prior to the placement or removal of artificial reefs or fish aggregation devices.

Article 9 : Aquaculture development

9.1 Responsible Development of Aquaculture, including Culture-based Fisheries, in Areas under National Jurisdiction

9.1.1 States should establish, maintain and develop an appropriate legal and administrative framework which facilitates the development of responsible aquaculture.

9.1.2 States should promote responsible development and management of aquaculture, including an advance evaluation of the effects of aquaculture development on genetic diversity and ecosystem integrity, based on the best available scientific information.

9.1.3 States should produce and regularly update aquaculture development strategies and plans, as required, to ensure that aquaculture development is ecologically sustainable and to allow the rational use of resources shared by aquaculture and other activities.

9.1.4 States should ensure that the livelihoods of local communities, and their access to fishing grounds, are not negatively affected by aquaculture developments.

9.1.5 States should establish effective procedures specific to aquaculture to undertake appropriate environmental assessment and monitoring with the aim of minimising adverse ecological changes and related economic and social consequences resulting from water extraction, land use, discharge of effluents, use of drugs and chemicals, and other aquaculture activities.

9.2 Responsible Development of Aquaculture including Culture-based fisheries within Transboundary Aquatic Ecosystems

9.2.1 States should protect transboundary aquatic ecosystems by supporting responsible aquaculture practices within their national jurisdiction and by cooperation in the promotion of sustainable aquaculture practices.

9.2.2 States should, with due respect to their neighbouring States, and in accordance with international law, ensure responsible choice of species, siting and management of aquaculture activities which could affect transboundary aquatic ecosystems.

9.2.3 States should consult with their neighbouring States, as appropriate, before introducing non-indigenous species into transboundary aquatic ecosystems.

9.2.4 States should establish appropriate mechanisms, such as databases and information networks to collect, share and disseminate data related to their aquaculture activities to facilitate cooperation on planning for aquaculture development at the national, subregional, regional and global level.

9.2.5 States should cooperate in the development of appropriate mechanisms, when required, to monitor the impacts of inputs used in aquaculture.

9.3 Use of Aquatic Genetic Resources for the Purposes of Aquaculture including Culture-based Fisheries

9.3.1 States should conserve genetic diversity and maintain integrity of aquatic communities and ecosystems by appropriate management. In particular, efforts should be undertaken to minimise the harmful effects of introducing non-native species or genetically altered stocks used for aquaculture including culture-based fisheries into waters, especially where there is a significant potential for the spread of such non-native species or genetically altered stocks into waters under the jurisdiction of other States as well as waters under the jurisdiction of the State of origin. States should, whenever possible, promote steps to minimise adverse genetic, disease and other effects of escaped farmed fish on wild stocks.

9.3.2 States should cooperate in the elaboration, adoption and implementation of international codes of practice and procedures for introductions and transfers of aquatic organisms.

9.3.3 States should, in order to minimise risks of disease transfer and other adverse effects on wild and cultured stocks, encourage adoption of appropriate practices in the genetic improvement of broodstocks, the introduction of non-native species, and in the production, sale and transport of eggs, larvae or fry, broodstock or other live materials. States should facilitate the preparation and implementation of appropriate national codes of practice and procedures to this effect.

9.3.4 States should promote the use of appropriate procedures for the selection of broodstock and the production of eggs, larvae and fry.

9.3.5 States should, where appropriate, promote research and, when feasible, the development of culture techniques for endangered species to protect, rehabilitate and enhance their stocks, taking into account the critical need to conserve genetic diversity of endangered species.

9.4 Responsible Aquaculture at the Production Level

9.4.1 States should promote responsible aquaculture practices in support of rural communities, producer organisations and fish farmers.

9.4.2 States should promote active participation of fishfarmers and their communities in the development of responsible aquaculture management practices.

9.4.3 States should promote efforts which improve selection and use of appropriate feeds, feed additives and fertilisers, including manures.

9.4.4 States should promote effective farm and fish health management practices favouring hygienic measures and vaccines. Safe, effective and minimal use of therapeutants, hormones and drugs, antibiotics and other disease control chemicals should be ensured.

9.4.5 States should regulate the use of chemical inputs in aquaculture which are hasardous to human health and the environment.

9.4.6 States should require that the disposal of wastes such as offal, sludge, dead or diseased fish, excess veterinary drugs and other hasardous chemical inputs does not constitute a hasard to human health and the environment.

9.4.7 States should ensure the food safety of aquaculture products and promote efforts which maintain product quality and improve their value through particular care before and during harvesting and on-site processing and in storage and transport of the products.

Article 10: Integration of fisheries into coastal area management

10.1 Institutional Framework

10.1.1 States should ensure that an appropriate policy, legal and institutional framework is adopted to achieve the sustainable and integrated use of the resources, taking into account the fragility of coastal ecosystems and the finite nature of their natural resources and the needs of coastal communities.

10.1.2 In view of the multiple uses of the coastal area, States should ensure that representatives of the fisheries sector and fishing communities are consulted in the decision-making processes and involved in other activities related to coastal area management planning and development.

10.1.3 States should develop, as appropriate, institutional and legal frameworks in order to determine the possible uses of coastal resources and to govern access to them taking into account the rights of coastal fishing communities and their customary practices to the extent compatible with sustainable development.

10.1.4 States should facilitate the adoption of fisheries practices that avoid conflict among fisheries resources users and between them and other users of the coastal area.

10.1.5 States should promote the establishment of procedures and mechanisms at the appropriate administrative level to settle conflicts which arise within the fisheries sector and between fisheries resource users and other users of the coastal area.

10.2 Policy Measures

10.2.1 States should promote the creation of public awareness of the need for the protection and management of coastal resources and the participation in the management process by those affected.

10.2.2 In order to assist decision-making on the allocation and use of coastal resources, States should promote the assessment of their respective value taking into account economic, social and cultural factors.

10.2.3 In setting policies for the management of coastal areas, States should take due account of the risks and uncertainties involved.

10.2.4 States, in accordance with their capacities, should establish or promote the establishment of systems to monitor the coastal environment as part of the coastal management process using physical, chemical, biological, economic and social parameters.

10.2.5 States should promote multi-disciplinary research in support of coastal area management, in particular on its environmental, biological, economic, social, legal and institutional aspects.

10.3 Regional Cooperation

10.3.1 States with neighbouring coastal areas should cooperate with one another to facilitate the sustainable use of coastal resources and the conservation of the environment.

10.3.2 In the case of activities that may have an adverse transboundary environmental effect on coastal areas, States should:

provide timely information and, if possible, prior notification to potentially affected States; and consult with those States as early as possible.

10.3.3 States should cooperate at the subregional and regional level in order to improve coastal area management.

10.4 Implementation

10.4.1 States should establish mechanisms for cooperation and coordination among national authorities involved in planning, development, conservation and management of coastal areas.

10.4.2 States should ensure that the authority or authorities representing the fisheries sector in the coastal management process have the appropriate technical capacities and financial resources.

Article 11: Post-harvest practices and trade

11.1 Responsible Fish Utilisation

11.1.1 States should adopt appropriate measures to ensure the right of consumers to safe, wholesome and unadulterated fish and fishery products.

11.1.2 States should establish and maintain effective national safety and quality assurance systems to protect consumer health and prevent commercial fraud.

11.1.3 States should set minimum standards for safety and quality assurance and make sure that these standards are effectively applied throughout the industry. They should promote the implementation of quality standards agreed within the context of the FAO/WHO Codex Alimentarius Commission and other relevant organisations or arrangements.

11.1.4 States should cooperate to achieve harmonisation, or mutual recognition, or both, of national sanitary measures and certification programmes as appropriate and explore possibilities for the establishment of mutually recognised control and certification agencies.

11.1.5 States should give due consideration to the economic and social role of the post-harvest fisheries sector when formulating national policies for the sustainable development and utilisation of fishery resources.

11.1.6 States and relevant organisations should sponsor research in fish technology and quality assurance and support projects to improve post-harvest handling of fish, taking into account the economic, social, environmental and nutritional impact of such projects.

11.1.7 States, noting the existence of different production methods, should through cooperation and by facilitating the development and transfer of appropriate technologies, ensure that processing, transporting and storage methods are environmentally sound.

11.1.8 States should encourage those involved in fish processing, distribution and marketing to:

— reduce post-harvest losses and waste;
— improve the use of by-catch to the extent that this is consistent with responsible fisheries management practices; and
— use the resources, especially water and energy, in particular wood, in an environmentally sound manner.

11.1.9 States should encourage the use of fish for human consumption and promote consumption of fish whenever appropriate.

11.1.10 States should cooperate in order to facilitate the production of value-added products by developing countries.

11.1.11 States should ensure that international and domestic trade in fish and fishery products accords with sound conservation and management practices through improving the identification of the origin of fish and fishery products traded.

11.1.12 States should ensure that environmental effects of post- harvest activities are considered in the development of related laws, regulations and policies without creating any market distortions.

11.2 Responsible International Trade

11.2.1 The provisions of this Code should be interpreted and applied in accordance with the principles, rights and obligations established in the World Trade Organisation (WTO) Agreement.

11.2.2 International trade in fish and fishery products should not compromise the sustainable development of fisheries and responsible utilisation of living aquatic resources.

11.2.3 States should ensure that measures affecting international trade in fish and fishery products are transparent, based, when applicable, on scientific evidence, and are in accordance with internationally agreed rules.

11.2.4 Fish trade measures adopted by States to protect human or animal life or health, the interests of consumers or the environment, should not be discriminatory and should be in accordance with internationally agreed trade rules, in particular the principles, rights and obligations established in the Agreement on the Application of Sanitary and Phytosanitary Measures and the Agreement on Technical Barriers to Trade of the WTO.

11.2.5 States should further liberalise trade in fish and fishery products and eliminate barriers and distortions to trade such as duties, quotas and non-tariff barriers in accordance with the principles, rights and obligations of the WTO Agreement.

11.2.6 States should not directly or indirectly create unnecessary or hidden barriers to trade which limit the consumer's freedom of choice of supplier or that restrict market access.

11.2.7 States should not condition access to markets to access to resources. This principle does not preclude the possibility of fishing agreements between States which include provisions referring to access to resources, trade and access to markets, transfer of technology, scientific research, training and other relevant elements.

11.2.8 States should not link access to markets to the purchase of specific technology or sale of other products.

11.2.9 States should cooperate in complying with relevant international agreements regulating trade in endangered species.

11.2.10 States should develop international agreements for trade in live specimens where there is a risk of environmental damage in importing or exporting States.

11.2.11 States should cooperate to promote adherence to, and effective implementation of relevant international standards for trade in fish and fishery products and living aquatic resource conservation.

11.2.12 States should not undermine conservation measures for living aquatic resources in order to gain trade or investment benefits.

11.2.13 States should cooperate to develop internationally acceptable rules or standards for trade in fish and fishery products in accordance with the principles, rights, and obligations established in the WTO Agreement.

11.2.14 States should cooperate with each other and actively participate in relevant regional and multilateral fora, such as the WTO, in order to ensure equitable, non-discriminatory trade in fish and fishery products as well as wide adherence to multilaterally agreed fishery conservation measures.

11.2.15 States, aid agencies, multilateral development banks and other relevant international organisations should ensure that their policies and practices related to the promotion of international fish trade and export production do not result in environmental degradation or adversely impact the nutritional rights and needs of people for whom fish is critical to their health and well being and for whom other comparable sources of food are not readily available or affordable.

11.3 Laws and Regulations Relating to Fish Trade

11.3.1 Laws, regulations and administrative procedures applicable to international trade in fish and fishery products should be transparent, as simple as possible, comprehensible and, when appropriate, based on scientific evidence.

11.3.2 States, in accordance with their national laws, should facilitate appropriate consultation with and participation of industry as well as environmental and consumer groups in the development and implementation of laws and regulations related to trade in fish and fishery products.

11.3.3 States should simplify their laws, regulations and administrative procedures applicable to trade in fish and fishery products without jeopardising their effectiveness.

11.3.4 When a State introduces changes to its legal requirements affecting trade in fish and fishery products with other States, sufficient information and time should be given to allow the States and producers affected to introduce, as appropriate, the changes needed in their processes and procedures. In this connection, consultation with affected States on the time frame for implementation of the changes would be desirable. Due consideration should be given to requests from developing countries for temporary derogations from obligations.

11.3.5 States should periodically review laws and regulations applicable to international trade in fish and fishery products in order to determine whether the conditions which gave rise to their introduction continue to exist.

11.3.6 States should harmonise as far as possible the standards applicable to international trade in fish and fishery products in accordance with relevant internationally recognised provisions.

11.3.7 States should collect, disseminate and exchange timely, accurate and pertinent statistical information on international trade in fish and fishery products through relevant national institutions and international organisations.

11.3.8 States should promptly notify interested States, WTO and other appropriate international organisations on the development of and changes to laws, regulations and administrative procedures applicable to international trade in fish and fishery products.

Article 12: Fisheries research

12.1 States should recognise that responsible fisheries requires the availability of a sound scientific basis to assist fisheries managers and other interested parties in making decisions. Therefore, States should ensure that appropriate research is conducted into all aspects of fisheries including biology, ecology, technology, environmental science, economics, social science, aquaculture and nutritional science. States should ensure the availability of research facilities and provide appropriate training, staffing and institution building to conduct the research, taking into account the special needs of developing countries.

12.2 States should establish an appropriate institutional framework to determine the applied research which is required and its proper use.

12.3 States should ensure that data generated by research are analysed, that the results of such analyses are published, respecting confidentiality where appropriate, and distributed in a timely and readily understood fashion,in order that the best scientific evidence is made available as a contribution to fisheries conservation, management and development. In the absence of adequate scientific information, appropriate research should be initiated as soon as possible.

12.4 States should collect reliable and accurate data which are required to assess the status of fisheries and ecosystems, including data on bycatch, discards and waste. Where appropriate, this data should be provided, at an appropriate time and level of aggregation, to relevant States and subregional, regional and global fisheries organisations.

12.5 States should be able to monitor and assess the state of the stocks under their jurisdiction, including the impacts of ecosystem changes resulting from fishing pressure, pollution or habitat alteration. They should also establish the research capacity necessary to assess the effects of climate or environment change on fish stocks and aquatic ecosystems.

12.6 States should support and strengthen national research capabilities to meet acknowledged scientific standards.

12.7 States, as appropriate in cooperation with relevant international organisations, should encourage research to ensure optimum utilisation of fishery resources and stimulate the research required to support national policies related to fish as food.

12.8 States should conduct research into, and monitor, human food supplies from aquatic sources and the environment from which they are taken and ensure that there is no adverse health impact on consumers. The results of such research should be made publicly available.

12.9 States should ensure that the economic, social, marketing and institutional aspects of fisheries are adequately researched and that comparable data are generated for ongoing monitoring, analysis and policy formulation.

12.10 States should carry out studies on the selectivity of fishing gear, the environmental impact of fishing gear on target species and on the behaviour of target and non-target species in relation to such fishing gear as an aid for management decisions and with a view to minimising non-utilised catches as well as safeguarding the biodiversity of ecosystems and the aquatic habitat.

12.11 States should ensure that before the commercial introduction of new types of gear, a scientific evaluation of their impact on the fisheries and ecosystems where they will be used should be undertaken. The effects of such gear introductions should be monitored.

12.12 States should investigate and document traditional fisheries knowledge and technologies, in particular those applied to small-scale fisheries, in order to assess their application to sustainable fisheries conservation, management and development.

12.13 States should promote the use of research results as a basis for the setting of management objectives, reference points and performance criteria, as well as for ensuring adequate linkages between applied research and fisheries management.

12.14 States conducting scientific research activities in waters under the jurisdiction of another State should ensure that their vessels comply with the laws and regulations of that State and international law.

12.15 States should promote the adoption of uniform guidelines governing fisheries research conducted on the high seas.

12.16 States should, where appropriate, support the establishment of mechanisms, including, inter alia, the adoption of uniform guidelines, to facilitate research at the subregional or regional level and should encourage the sharing of the results of such research with other regions.

12.17 States, either directly or with the support of relevant international organisations, should develop collaborative technical and research programmes to improve understanding of the biology, environment and status of transboundary aquatic stocks.

12.18 States and relevant international organisations should promote and enhance the research capacities of developing countries, inter alia, in the areas of data collection and analysis, information, science and technology, human resource development and provision of research facilities, in order for them to participate effectively in the conservation, management and sustainable use of living aquatic resources.

12.19 Competent international organisations should, where appropriate, render technical and financial support to States upon request and when engaged in research investigations aimed at evaluating stocks which have been previously unfished or very lightly fished.

12.20 Relevant technical and financial international organisations should, upon request, support States in their research efforts, devoting special attention to developing countries, in particular the least-developed among them and small island developing countries.

Annex 1. Background to the Origin and Elaboration of the Code

1. This annex describes the process of elaboration and negotiation of the Code, which led to its submission for adoption to the Twenty-eighth Session of the FAO Conference. It has been felt useful to annex this section as a reference to the origin and the development of the Code and thus reflect the interest generated and the spirit of compromise of all the parties involved in its elaboration. It is hoped that this will contribute to the promotion of the commitment necessary for its implementation.

2. At various international fora, concern had long been expressed regarding the clear signs of over-exploitation of important fish stocks, damage to ecosystems, economic losses, and issues affecting fish trade - all of which threatened the long-term sustainability of fisheries and, in turn, harmed the contribution of fisheries to food supply. In discussing the current state and prospects of world fisheries, the Nineteenth Session of the FAO Committee on Fisheries (COFI), held in March 1991, recommended that FAO should develop the concept of responsible fisheries and elaborate a Code of Conduct to this end.

3. Subsequently, the Government of Mexico, in collaboration wiht FAO, organised an International Conference on Responsible Fishing in Cancûn, in May 1992. The Declaration of Cancûn endorsed at that Conference further developed the concept of responsible fisheries, stating that "this concept encompasses the sustainable utilisation of fisheries resources in harmony with the environment; the use of capture and aquaculture practices which are not harmful to ecosystems, resources or their quality; the incorporation of added value to such products through transformation processes meeting the required sanitary standards; the conduct

of commercial practices so as to provide consumers access to good quality products".

4. The Cancûn Declaration was brought to the attention of the UNCED Rio Summit in June 1992, which supported the preparation of a Code of Conduct for Responsible Fisheries. The FAO Technical Consultation on High Seas Fishing, held in September 1992, further recommended the elaboration of a Code to address the issues regarding high seas fisheries.

5. The One Hundred and Second Session of the FAO Council, held in November 1992, discussed the elaboration of the Code, recommending that priority be given to high seas issues and requested that proposals for the Code be presented to the 1993 session of the Committee on Fisheries.

6. The Twentieth Session of COFI, held in March 1993, examined general principles for such a Code, including the elaboration of guidelines and endorsed a timeframe for the further elaboration of the Code. It also requested FAO to prepare, on a "fast track" basis, as part of the Code, proposals to prevent reflagging of fishing vessels which affect conservation and management measures on the high seas.

7. The further development of the Code of Conduct for Responsible Fisheries was accordingly carried out in consultation and collaboration with relevant United Nations Agencies and other international organisations including non-governmental organisations.

8. In pursuance of the instructions of the FAO Governing Bodies, the draft Code was formulated in such a way as to be consistent with the 1982 United Nations Convention on the Law of the Sea, taking into account the 1992 Declaration of Cancûn, the 1992 Rio Declaration and the provisions of Agenda 21 of UNCED, the conclusions and recommendations of the 1992 FAO Technical Consultation on High Seas Fishing, the Strategy endorsed by the 1984 FAO World Conference on Fisheries Management and Development, and other relevant instruments including the outcome of the then ongoing United Nations Conference on Straddling Fish Stocks and Highly Migratory Fish Stocks which, in August 1995, adopted an Agreement for the Implementation of the Provisions of the United Nations Convention on the Law of the Sea of 10 December 1982 Concerning Straddling Fish Stocks and Highly Migratory Fish Stocks.

9. The FAO Conference, at its Twenty-seventh Session in November 1993, adopted the Agreement to Promote Compliance with International

Conservationa and Management Measures by Fishing Vessels on the High Seas and recommended that the General Principles of the Code of Conduct for Responsaible Fisheries be prepared on a "fast track" in order to orientate formulation of thematic articles. Accordingly, a draft text of the General Principles was reviewed by an informal Working Group of Government-nominated experts, which met in Rome in February 1994. A revised draft was widely circulated to all FAO Members and Associate Members as well as intergovernmental and non-governmental organisations. Comments received on the second version of the General Principles were incorporated in the draft Code together with proposals for an alternative text. This document was also the subject of informal consultation with non-governmental organisations on the occasion of the Fourth Session of the United Nations Conference on Straddling Fish Stocks and Highly Migratory Fish Stocks, held in August 1994 in New York.

10. In order to facilitate consideration of the full text of the draft Code, the Director-General proposed to the Council at its Hundred and Sixth Session inJune 1994, that a Technical Consultation on the Code of Conduct for Responsible Fishing be organised, open to all FAO Members, interested non-members, intergovernmental and non-govermental organisations, in order to provide an opportunity for the widest involvement of all concerned parties at an early stage of its elaboration.

11. This Technical Consultation took place in Rome from 26 September to 5 October 1994 and a draft for the entire Code and a first draft of technical guidelines to support most of the Thematic Articles of the Code were presented. Following a thorough review of all the Articles of the complete draft Code of Conduct, an Alternative Secretariat Draft was then prepared on the basis of comments made during the discussions in plenary and specific drafting changes submitted in writing during the Consultation.

12. The Consultation was able to review also in detail an alternative draft for three of the six Thematic Articles of the Code, i.e., Article 9 "Integration of Fisheries into Coastal Area Management", Article 6 "Fisheries Management", Article 7 "Fishing Operations", except for those principles which were likely to be affected by the outcome of the ongoing UN Conference on Straddling Fish Stocks and Highly Migratory Fish Stocks. A short Administrative Report was prepared and presented to the FAO Council and to COFI.

13. The Technical Consultation proposed to the Council at its Hundred and Seventh Session, 15-24 November 1994, that the final wording of those principles dealing mainly with high seas issues be left in abeyance pending the outcome of the UN Conference. The Council generally endorsed the proposed procedure, noting that following discussions at the next session of COFI, a final draft of the Code would be submitted to the FAO Council in June 1995 which would then decide upon the necessity for a Technical Committee to meet in parallel to that Session of the Council in order to elaborate further the detailed provisions of the Code if required.

14. Based upon the substantial comments and detailed suggestions received at the Technical Consultation, the Secretariat elaborated a revised draft of the Code of Conduct for Responsible Fisheries, which was submitted to the Twenty-first Session of the Committee on Fisheries, held from 10 to 15 March 1995.

15. The Committee on Fisheries was also informed that the UN Conference was expected to conclude its work in August 1995. It was proposed that principles left in abeyance in the draft text of the Code could then be reconciled with the language agreed upon at the UN Conference in accordance with a mechanism to be decided upon by the Committee and the Council, beforesubmission of the complete Code for its adoptiton at the Twenty-eighth Session of the FAO Conference in October 1995.

16. The Committee was informed of the various steps the Secretariat had undertaken in preparing the draft Code of Conduct. The Committee established an open-ended Working Group in order to review the draft text of the Code. The Working Group, which met from 10 to 14 March 1995, undcrtook a detailed revision of the draft Code in continuation of the work carried out by the Technical Consultation. It completed and approved the text of Articles 8 to 11. In view of the time constraints, the Working Group provided directives to the Secretariat to redraft Articles 1 to 5. It was also recommended that the elements of research and cooperation as well as aquaculture be included in Article 5, General Principles, to reflect issues developed in the Thematic Articles of the Code.

17. The Committee supported the proposal endorsed by the Hundred and Seventh Session of the Council on mechanisms to finalise the Code. The final wording of those principles dealing mainly with issues concerning straddling fish stocks and highly migratory fish stocks, which formed only a small part of the Code, should be re-examined in the light of the outcome

of the UN Conference. The Group also recommended that once agreement was reached on the substance, it would be necessary to harmonise legal, technical and idiomatic aspects of the Code, in order to facilitate its final approval.

18. The Report of the open-ended Working Group was presented to a Ministerial Meeting on Fisheries, held on 14 and 15 March 1995, in conjunction with the COFI Session. The Rome Consensus on World Fisheries emanating from this meeting urged that "Governments and international organisations take prompt action to complete the International Code of Conduct for Responsible Fisheries with a view to submitting the final text to the FAO Conference in October 1995".

19. The Hundred and Eighth Session of the Council was presented with a revised version of the Code of Conduct. The Council established an open-ended Technical Committee, which held its First Session from 5 to 9 June 1995, with a broad regional representation of members and observers. A number of intergovernmental and non-governmental organisations also participated.

20. The Council was informed by the Technical Committee that it had undertaken a thorough review of Articles 1 to 5 including the Introduction. It had also examined, amended and approved Aticles 8 to 11. The Council was also informed that the Committee had started the revision of Article 6.

21. The Council approved the work carried out by the Technical Committee and endorsed its recommendation for a Second Session to be held from 25 to 29 September 1995 to complete the revision of the Code once the Secretariat had harmonised the text linguistically and juridically, taking into account the outcome of the UN Conference on Straddling Fish Stocks and Highly Migratory Fish Stocks.

22. A revised version of the Code as approved by the Open-ended Technical Committee at its First Session (5-9 June 1995) and endorsed by the One Hundred and Eighth Session of the Council was issued, both as a Conference document (C 95/20) and as a working paper for the Second Session of the Technical Committee. Elements pending agreement were clearly identified.

23. In order to facilitate the finalisation of the entire Code, the Secretariat prepared the document "Secretariat Proposals for Article 6, Fisheries Management, and Article 7, Fishing Operations, of the Code of

Conduct for Responsible Fisheries", taking into account the Agreement relating to the Conservation and Management of Straddling Fish Stocks and Highly Migratory Fish Stocks, adopted by the UN Conference in August 1995. The Secretariat also completed proposals for the harmonisation of the text on legal and linguistic aspects and made this available to the Committee in three languages for the session (English, French and Spanish).

24. A Second Session of the Open-ended Technical Committee of the Council met from 25 to 29 September 1995, with a wide representation of regions and interested organisations. The Committee, working in a full spirit of collaboration, successfully concluded its mandate, finalising and endorsing all Articles and the Code as a whole. The Technical Committee agreed that the negotiations of the text of the Code were finalised. An Open-ended Informal Group on Language Harmonisation held an additional session and, together with the Secretariat, completed the harmonisation on the basis of the text as adopted at the closing session. The Technical Committee instructed the Secretariat to already submit the finalised version as a revised Conference document to the Hundred and Ninth Session of the Council and to the Twenty-eighth Session of the Conference for its adoption. The Council endorsed the Code of Conduct as finalised by the Technical Committee. The Secretariat was requested to prepare the required draft resolution for the Conference, including also a call on countries to ratify, as a matter of urgency, the Compliance Agreement adopted at the last session of the Conference. The Twenty-eighth Session of the Conference adopted on 31 October 1995, by consensus, the Code of Conduct for Responsible Fisheries and the respective Resolution.

Annex 2 . Resolution

The Conference,

Recognising the vital role of fisheries in world food security, and economic and social development, as well as the need to ensure the sustainability of the living aquatic resources and their environment for present and future generations,

Recalling that the Committee on Fisheries on 19 March 1991 recommended the development of the concept of responsible fishing and the possible formulation of an instrument on the matter,

Considering that the Declaration of Cancún, which emanated from the Inter-national Conference on Responsible Fisheries of May 1992, organised by the Government of Mexico in collaboration with FAO, had called for the preparation of a Code of Conduct on Responsible Fisheries.

Bearing in mind that with the entry into force of the United Nations Convention on the Law of the Sea, 1982, and the adoption of the Agreement for the implementation of the provisions of the United Nations Convention on the Law of the Sea of 10 December 1982 Relating to the Conservation and Management of Straddling Fish Stocks and Highly Migratory Fish Stocks, as was anticipated in the 1992 Rio Declaration and the provisions of Agenda 21 of UNCED, there is an increased need for subregional and regional cooperation, and that significant responsibilities are placed upon FAO in accordance with its mandate,

Recalling further that the Conference in 1993 adopted the FAO Agreement to Promote Compliance with International Conservation and Management Measures by Fishing Vessels on the High Seas, and that this Agreement would constitute an integral part of the Code of Conduct,

Noting with satisfaction that FAO, in accordance with the decisions of its Governing Bodies, had organised a series of technical meetings to formulate the Code of Conduct, and that these meetings have resulted in agreement being reached on the text of the Code of Conduct for Responsible Fisheries,

Acknowledging that the Rome Consensus on World Fisheries, which emanated from the Ministerial Meeting on Fisheries of 14-15 March 1995, urged govern-ments and international organisations to respond effectively to the current fisheries situation, inter alia, by completing the Code of Conduct for Responsible Fisheries and to consider adopting the Agreement to Promote Compliance with International Conservation and Management Measures by Fishing Vessels on the High Seas:

1. Decides to adopt the Code of Conduct for Responsible Fisheries;
2. Calls on States, International Organisations, whether Governmental or Non-Governmental, and all those involved in fisheries to collaborate in the fulfilment and implementation of the objectives and principles contained in this Code;
3. Urges that special requirements of developing countries be taken into account in implementing the provisions of this Code;
4. Requests FAO to make provision in the Programme of Work and Budget for providing advice to developing countries in implementing

this Code and for the elaboration of an Interregional Assistance Programme for external assistance aimed at supporting implementation of the Code;

5. Further requests FAO, in collaboration with members and interested relevant organisations, to elaborate, as appropriate, technical guidelines in support of the implementaiton of the Code;
6. Calls upon FAO to monitor and report on the implementation of the Code and its effects on fisheries, including action taken under other instruments and resolutions by UN organisations and, in particular, the resolutions adopted by the General Assembly to give effect to the Conference on Straddling Fish Stocks and Highly Migratory Fish Stocks leading to the Agreement for the implementation of the provisions of the United Nations Convention on the Law of the Sea of 10 December 1982 Relating to the Conservation and Management of Straddling Fish Stocks and Highly Migratory Fish Stocks;
7. Urges FAO to strengthen Regional Fisheries Bodies in order to deal more effectively with fisheries conservation and management issues in support of subregional, regional and global cooperation and coordination in fisheries.

Bibliography

Adams, T., "Coastal fisheries and marine development issues for small islands", In M.J. M. J. Williams, ed., *A roadmap for the future for fisheries and conservation, ICLARM Conference Proceedings* No. 56: 40 - 50, 1998.

Allain, R.J. 1988. Monitoring, control and surveillance in selected West African coastal states with recommendations for possible ICOD interventions in the future. Ottawa, ICOD.

Arlinghaus, R. 2006. Overcoming human obstacles to conservation of recreational fishery resources, with emphasis on central Europe. *Environmental Conservation*, 33: 46–59.

Bell, J.D., "Transfer of technology on marine ranching to small island states", *In (unedited) Marine ranching: global perspectives with emphasis on the Japanese experience, FAO Fisheries Circular no.*, 943: 53 - 65, 1999b.

Bennett, E., Valette, H.R., Mäiga, K.Y. and Medard, M., eds. 2004. *Room to manoeuvre: gender and coping strategies in the fisheries sector.* Portsmouth, UK, IDDRA. 154 pp

Bergin A. 1988. Fisheries surveillance in the South Pacific. *Ocean and Coastal Management*, 1988, Vol.11, No.6. pp. 467–491. Australia.

Bonell A.D. 1994. *Quality assurance in seafood processing; A practical guide.* Chapman & Hall, New York, London.

Charles, A.T., Use rights and responsible fisheries: limiting access and harvesting through rights-based management, in *A fishery manager's guidebook – Management measures and their application*, K.L. Cochrane (ed.), FAO Fisheries Technical Paper, No. 424, pp. 131–157.

Clark, J.R. 1992. Integrated management of coastal zones. *FAO Fisheries Technical Paper* 327. Rome, FAO.

Commonwealth of Australia, *National strategy for ecologically sustainable development,* Canberra, Australian Government Publishing Service, 1992.

Dalzell, P., Adams, T.J.H. & Polunin, N.V.C., "Coastal fisheries in the Pacific Islands", *Oceanogr. Mar. Biol.: an Annual Review*, 34: 395 - 531, 1996.

FAO Fisheries Resources Division, I*ndicators for sustainable development of marine capture fisheries,* FAO Technical Guidelines for Responsible Fisheries, No. 8, 1999.

FAO. 1992. Monitoring, control and surveillance of fisheries in the exclusive economic zones of Asean countries: project findings and recommendations, *MCS training, completed manuals and course materials*. FAO FI:DPRAS/86/115. Rome/ Jakarta, FAO.

FAO. 1995. Code of conduct for responsible fisheries. Rome, FAO, 41 pp.

FAO. 2010. *Aquaculture development. 4. Ecosystem approach to aquaculture.* FAO Technical Guidelines for Responsible Fisheries. No. 5, Suppl. 4. Rome. 53 pp.

Garcia, S.M. & Grainger, R.J.R. 2005. Gloom and doom? The future of marine capture fisheries. *Philosophical transactions of the Royal Society of London – Series B: biological sciences.*

Gascoigne, J. and Willsteed, E. 2009. *Moving towards low impact fisheries in Europe: policy hurdles & actions.* Brussels, Seas At Risk. 103 pp.

Jakobsen M., Lillie A. 1992. "Quality systems for the fish industry". In *Quality assurance in the fish industry*. Eds: H.H. Huss, M. Jakobsen, J. Liston.

Jentoft, S. 2006. Beyond fisheries management: the phronetic dimension. *Marine Policy,* 30(6): 671- 680.

MacCall, A.D., "Against marine fish hatcheries: ironies of fishery politics in the technological era", *Cal. COFI Reports*, 30: 46 - 48, 1989.

McCarty, C.E., McEachron, L.W. & Rutledge,W.P., "Beneficial uses of marine fish hatcheries: the Texas experience", *The International Symposium on Sea Ranching of Cod and Other Marine Species*, Arendal, Norway, 15 - 18 June 1993, Programme and Abstracts, 1993.

Miller, K.A. 2007. Climate variability and tropical tuna: management challenges for highly migratory fish stocks. *Marine Policy.*

Morton, J.F. 2007. The impact of climate change on smallholder and subsistence agriculture. *Proceedings of the National Academy of Sciences.*

Newton, C.H. 1989. Monitoring, control and surveillance of fisheries in exclusive economic zones. *RSC Series* No. 49. pp. 209-213. Rome, FAO.

Sinha, V.R.P. 1996. "Shrimp Culture, Trade and the Environment in India". *INFOFISH*, Kualalumpur (Mimeo), 31p.

World Wildlife Foundation Australia, *Policy proposals and operational guidance for ecosystem-based management of marine capture fisheries,* 2002